PARTICIPATORY DEVELOPMENT IN KENYA

Voices in Development Management

Series Editor:
Margaret Grieco
Napier University, Scotland

The Voices in Development Management series provides a forum in which grass roots organisations and development practitioners can voice their views and present their perspectives along with the conventional development experts. Many of the volumes in the series will contain explicit debates between various voices in development and permit the suite of neglected development issues such as gender and transport or the microcredit needs of low income communities to receive appropriate public and professional attention.

Also in the series

Dot Com Mantra
Social Computing in the Central Himalayas
Payal Arora
ISBN 978 1 4094 0107 0

The Dominance of Management
A Participatory Critique
Leonard Holmes
ISBN 978 0 7546 1184 4

Losing Paradise
The Water Crisis in the Mediterranean
Edited by Gail Holst-Warhaft and Tammo Steenhuis
ISBN 978 0 7546 7573 0

Tourism, Development and Terrorism in Bali
Michael Hitchcock and I Nyoman Darma Putra
ISBN 978 0 7546 4866 6

Women Miners in Developing Countries
Pit Women and Others
Edited by Kuntala Lahiri-Dutt and Martha Macintyre
ISBN 978 0 7546 4650 1

Participatory Development in Kenya

JOSEPHINE SYOKAU MWANZIA
Ministry of State for Public Service, Kenya
ROBERT CRAIG STRATHDEE
Victoria University of Wellington, New Zealand

ASHGATE

Published by
Ashgate Publishing Limited
Wey Court East
Union Road
Farnham
Surrey, GU9 7PT
England

Ashgate Publishing Company
Suite 420
101 Cherry Street
Burlington
VT 05401-4405
USA

www.ashgate.com

British Library Cataloguing in Publication Data
Mwanzia, Josephine Syokau.
Participatory development in Kenya. -- (Voices in development management)
1. Community development--Kenya. 2. Fundamental education--Kenya. 3. Educational planning--Kenya--Citizen participation--Case studies.
I. Title II. Series III. Strathdee, Robert.
361.2'5'096762-dc22

Library of Congress Cataloging-in-Publication Data
Mwanzia, Josephine Syokau.
Participatory development in Kenya / Josephine Syokau Mwanzia and Robert Craig Strathdee.
p. cm. -- (Voices in development management)
Includes bibliographical references and index.
ISBN 978-0-7546-7877-9 (hardback) -- ISBN 978-0-7546-9732-9 (ebook)
1. Community development--Kenya--Citizen participation. I. Strathdee, Robert.
II. Title.

HN793.Z9C6492 2010
307.1'4096762--dc22

2010017416

ISBN 978 0 7546 7877 9 (hbk)
ISBN 978 0 7546 9732 9 (ebk)

Printed and bound in Great Britain by
MPG Books Group, UK

Contents

List of Figures and Table

Figures

Table

Foreword

Turning Weakness, Oppression and Despair into Opportunity and Strength

In recent years there has been great revival of interest in participatory development, gender equality, human capital development and good governance, not only as urgent but politically hot topics. It is not difficult to see why. Apparently, Africa is falling behind the rest of the world in spite of its vast human and natural resources potential,s. The underlying yet crucial and endemic problem the continent faces is the inability to turn weaknesses into opportunities; and threats into strengths for the common good of the people.

Unfortunately, we turn to the "blame theory" to hide our weaknesses and failures in constructing a developmental state, which at least, guarantees a satisfactory quality of life for a reasonably sizeable proportion of its citizens. The neglect of the female gender due to archaic customs and male chauvinism is holding back the sustainable socioeconomic transformation of the continent. Women who, in many countries constitute more than 52 per cent of the population, are marginalized and not included in the mainstream development and decision-making processes. "Women hold up half the sky" and they endeavour to prove the accuracy of this statement.

Think about the major issues confronting the African continent this century. These include war, ethnic diversity and conflicts, insecurity and terrorism; population pressures, environmental strains and climate change, poor human capital development; inappropriate utilization of natural resources, poverty and income gaps. For all these diverse problems, empowering women is part of the answer. Most obviously, educating girls and bringing them into the formal economy will yield economic dividends and help address national and global poverty. Thus the education of girls and the provision of microfinance for rural and urban women can pay huge dividends in terms of ending oppression and creating opportunities for development.

Educating girls is one of the most effective ways of fighting poverty, marginalization and underdevelopment. The authors argue that one way to soothe some conflict-ridden or underdeveloped societies like Kenya is to bring women and girls into schools, the workplace, government, markets, and business partly to boost the economy and partly to ease the testosterone-laden values of African societies. Building the human capital base and ensuring a healthy society is key to improving production and giving added value to the natural resources base of the country.

Education is considered as one of those achievements which translate into change. Education alone may not produce change if other factors, or the will for development is not accelerated in other social, economic and political areas. Development must create employment and enable people to participate in the production of goods and quality services delivery. Knowledge is power. Not being armed with knowledge, the people will continue to swim in darkness and constantly be exploited by those who have a mastery of knowledge, information, science and technology as inputs to the transformation process.

Participation and development are part of the basic tenets of democracy. Participation is both a right and an empowering instrument for the people to gain control over life-enhancing systems and structures. According to Oxfam, participation is a fundamental right. It is a means of engaging poor people in joint analysis and development of priorities. Its ultimate goal should be to foster the existing capacities of poor, local women and men and to increase their self-reliance in ways that outlast specific projects. The purpose of participation is to give a permanent voice to poor or marginalized people and integrate them into mainstream decision-making structures and processes that shape their lives and the destiny of their society. The issue of social democracy stands tall, emphasizing political freedoms, rights and serious investment in the educational and health sector to enable people to participate in Kenyan society in a meaningful way. Affirmative action policies have been enacted to support education and healthcare. However, to allow poor people's meaningful participation, the physical and social infrastructures (learning institutions, healthcare facilities, transport and markets) that support sustainable livelihoods, must be developed throughout the country.

While this has been the focus, development has been top-down, ethnocentric, technocratic and based on Western ideas and structures barely taking into account time and space influenced by advances in science and technology. Africa does not need to re-invent the wheel, implying that the Rostowian "stages of development" is archaic and constitutes a Western, hidden agenda to delay the socioeconomic transformation of Africa into a modern society.

What the continent should do is skilfully exploit "alternatives" and aim for visionary servant leadership by incorporating all components of society in the development process. In short – "bottom-up" development is advocated as the way forward. The dichotomy between the state and the people must disappear to make room for collaborative and participatory roles for every Kenyan in the nation- building process.

This book offers an opportunity to reflect on the issue of state construction in Africa; why good governance, although not the panacea, provides the best opportunity for the continent to take its rightful place within the global community. What form of state construct did the departing colonial masters bequeath their colonies? What have Africans done to ensure that this state construct responds to the needs of their population? How do the people participate in the existing state structure? Why have we allowed "exclusion" to take precedence over "inclusion"?

These and many other pertinent questions underpin the focus of *Participatory Development in Kenya*, which can be summed up as: the search for a sustainable quality livelihood for the citizens, and their holistic involvement in crafting their destiny, developing and utilizing their resources in the first place, and receiving whatever external assistance is available as part of the global community in the process of crafting a new brave society. A society where the rule of law reigns, issues of ethical and moral values, accountability and transparency, honesty, responsibility and obligations, and empowerment are vigorously observed.

Mindful of this, civil society in Kenya, as in many sub-Saharan African countries, does not sufficiently exercise these attributes to push for democratic change which should ensure participation and development. Good governance is lacking in Kenya's development. Thus strong civil pressure is the missing link for better governance in Kenya and the rest of the continent. Civil society is weak, passive, disorganized, infiltrated and co-opted by the state, and beset with ethnic and civil strife. It is argued that participation has a greater chance than representative governance of ensuring effectiveness and efficiency in development; of promoting transparency and sustainability of development; of empowering people and breaking the chains of dependence; avoiding the ghettoes of poverty and underdevelopment; of promoting accountability and responsiveness to local needs, and of reducing the huge inequalities between the rich elite and the suffering, poor, silent majority.

This book is about Kenya, yet it depicts the common trend over the whole African continent. A trend signifying that the euphoria that accompanied the independence of the continent has gradually given way to rising despair and destructive frustration as efforts at development and nation-building have failed. The people feel betrayed and sidelined in the development process. Poor leadership, passiveness in civil society, an international community hostile to the needs and aspirations of the people has, among other things, paved the way for "irresponsible experiments which have collapsed indigenous knowledge, infrastructure, and cultural and social networks upon which the poor depend for survival". The publication digs deep into the Basic Education Improvement Project (BEIP) or human capital development "designed to give every Kenyan the right to quality education and training no matter their socio-economic status". This book goes further to underscore the role and place of democratic governance in accelerating societal changes.

Democratic governance is one of the weakest links in African development. Democratic governance must be homegrown without bypassing the universally accepted basic tenets or ingredients of democracy – respect for fundamental human rights and the inalienable rights of the people constructing their future. With this universally accepted creed, there are alternatives to development as the pathway towards the deconstruction of the wrong state construct inherited from colonial powers. *Participatory Development in Kenya* provides a valuable platform forward for constructive engagement of the citizens in taking their destiny into their hands.

While participatory democracy advances many potential benefits for the marginalized poor, there are questions as to how participatory governance and development could be achieved in practice in Kenya and tropical African states that are latecomers in the democracy race. Thus, this well-articulated book addresses the puzzle by examining the role of associations and associational life in the push for a deepened democracy and sustainable development at all levels of Kenyan society. Participatory development and governance hinges on associational life, the existence of shared norms, networks and solidarity among people that promotes collaborative action for perceived common good.

This network of social ties is commonly referred to as social capital and is generally defined as the glue that holds groups and societies together and forms bonds of shared values, norms and institutions. As pointed out in many sections of the book, the gluing and bridging of informal communities, such as community networks or formally-structured groups, to create strong pressure for greater participation in Kenya is limited as is the case with civil associations. The duty of researchers is to analyse how institutional factors can push the emerging participatory governance and development into crisis and how changes in rules, customs and practices can counteract this trend.

It is left for Kenyans to take the bold step in crafting a governance structure mainstreaming participatory development and enshrining the cardinal values that addresses the structural rights and challenges of disadvantaged and marginalized people. Populations continue to fail to attain emancipation, sustainable development and high quality standards of living because existing state construction relegates the importance of equalities caused by eclecticism, and political party influenced bureaucratic culture, that promotes centralized ethnic homogeneity cultural construct, gender inequality, abject poverty, destructive ethnicity structures and capitalist economies without a human face.

Participatory Development must be built on the foundations of democratic governance and powers must be shared to enforce the concept and realization of "participation" and the services of development for the common good. The structure of the book is one that reflects key issues related to the potential for and challenges of constructing a developmental state in Kenya that espouses issues of open and shared government; shared obligations and responsibilities; inclusion, not exclusion, and shared prosperity. The socioeconomic and cultural diversity of Africa means that the 'one-size fits all' approach is flawed. But common characteristics providing generalization exist. In this respect, issues of servant visionary leadership, structural changes in mindset, attitude and behaviour of the people, governance form and responses of the international community towards the needs and aspirations of the people provide the background to advancing a common approach and solutions to the predicaments of the people.

Captivating individual chapters explore, explain and analyse state construction, where things have gone wrong and what is to be done to ensure that Kenya registers itself as a worthy member of the community of nations using development, participation and empowerment theories. The methodological approach adopted puts as a focal point the issues of "top-down" and "bottom-up" pressures, associations and associational life. This strengthens the conceptual framework of the book. The research takes a one-country-analysis of how institutions affect associational life and 'bottom-up' pressure for participatory governance in order to control wide varieties of institutions. Kenya is a suitable choice because it is typical case of a gradually-stalled democracy in Africa judging from the events after the 2007–2008 election.

In addition, this book contains useful tables, figures and case studies including chapter outlines and conclusions specific to Kenya. It provides a wide range of relevant and important ideas, which must be examined in depth – even debated rigorously at a time when social tension, conflicts, unrests, ethnic division and the politics of exclusion are undergoing significant changes. It touches on issues raised at the "Africa 21 Conference", 18–19 May 2010 in Yaoundé, Cameroon, calling for democratic governance, greater political expression and access to a wider citizenry by bridging political divides and encouraging dialogue and tolerance. This would create a forum for nurturing capable and accountable political leaders, disseminating information for transparency and popular mobilization, and increasing the legitimacy of governance. The proper and efficient mobilization, training and utilization of human and natural resources could transform the continent. The sustainable transformation of Kenya is the message being sent home through this book.

It creates room for grassroots engagement in searching for alternatives to mainstream development practices, which alienate and degrade people's knowledge and culture as essential inputs. Importantly, Participatory Development puts an end to centralized authoritarian governance rule and "sit-tight" presidents for life which Africa is known for. This implies that democratic, cooperative and peaceful, rather than non-democratic, competitive and violent ways of problem-solving are better placed to emancipate and achieve sustainable development.

Kenya and the rest of Africa can draw useful example from the peaceful resolution of the Bakassi Peninsular conflict between Cameroon and Nigeria. It shows that only Africans have the right solutions to their problems. *Participatory Development in Kenya* is the foretaste to the reality that only Kenyans have the solution to the problems undermining its socioeconomic transformation. Has the BEIP created centres of excellence for centralization or decentralization? Or has it created and reinforced participatory and representative democracy, consensus-building or authoritarian rule.

It is apparent that the authors seek to maintain openness and share knowledge with the rest of the academic world. Through this book they provide a forum where different currents meet and cross-fertilize ideas in the ultimate best interest for society to move forward. The book offers reading materials and entrusts readers with judgement and critique of state construction, leadership, role of citizens, output functions of the state, the private sector and responses from the international community. The sad events that rocked Kenya during the post-elections of 2007 and ongoing conflicts in many parts of the continent call for serious soul searching on the development path embarked upon by African countries currently languishing in abject poverty in the midst of plenty.

This book, coming after the horrendous events in Kenya (2007), strongly emphasizes the "politics of inclusion and participatory development" as the foundation for growth, progress and quality living standards. The book drives home the single message to the Kenyan as well as to the African population to constructively engage in the politics of "inclusion and not exclusion" as the lynch pin for empowerment, security, development, respect for human rights and rule of law. Ethnic diversity and exclusion that contributed to the sad events following the December 2007 elections should be serious lessons for Kenya and other African countries in building bridges across the existing divide; and making ethnic diversity an asset, not a liability, to development.

Kenyans and the rest of the continent need deliberative democracy as an essential component of participatory democracy. This is a thoughtful contribution to the development and state construct discourses of the 21st century; and the debate on the relationship between state and citizens offering echoes ringing beyond the frontiers of Kenya. Above all, it puts to test the need to make the developmental state a centerpiece for the socioeconomic transformation of African states. This book should be the torch-bearer for kick-starting participatory development and democratic governance on the continent.

The role of incorporating informal and formal networks in "bottom-up" mobilization cannot be overemphasized. Such networks are imperative for propelling participatory development and democratic governance in Kenya and throughout the continent. There has to be a new dynamic of power and politics of inclusion rather than exclusion to take Kenya forward into the 21st century.

If there is a shortcoming with the book it is simply a question of arousing the interests of readers. Should the public not, after being served with such a delicious breakfast, expect lunch and dinner, if not late-night coffee? That is the question. *Participatory Development in Kenya* is not just a book to read and replace on the bookshelf, but one requiring a focused action-oriented inward looking strategy of lighting the burning flame of inclusion, constructive engagement, development and equity, shared-prosperity and injecting in the minds of the people concerted and comprehensive ideas geared towards meaningful changes in the daily lives of the Kenyan population. This is a book which makes a valuable contribution to knowledge and cross-fertilization of ideas for the good of society and humanity and should be given the widest possible publicity.

John W. Forje, Fil dr
Centre for Action-Oriented Research on African Development, CARAD
Post Box 13429 Yaoundé, Cameroon
and
Visiting Lecturer,
Department of Political Science,
University of Buea, Cameroon

Acknowledgements

The ideas recorded in this book were conceived in 2005, a time when the education sector in Kenya was receiving hefty amounts of aid to support universal primary and continuing education from the World Bank, the Department for International Development (DFID) and other international development agencies.

It was instructive to ask the question on the extent to which aid projects within the education sector, in a real sense, empowered and socially transformed the disadvantaged communities who participated in them.

Our thanks go to the innumerable participants who contributed ideas to answer this question. Educationists, teachers, parents and members of Boards of Governors, School Management Committees and Parent–Teacher Associations, your openness made our discussions quite captivating and the subjects addressed more meaningful. We are highly indebted to you for the time you spent conversing with us. Thank you for your support and for trusting us to be the "instruments" through which to raise your voices.

List of Abbreviations

APRM	African Peer Review Mechanism
ASALs	Arid and Semi-arid Lands
BEIP	Basic Education Improvement Project (or the Project)
BOG	Board of Governors
CBO	Community Based Organization
DFID	Department for International Development
FGM	Female Genital Mutilation
GOK	Government of Kenya
IMF	International Monetary Fund
NGO	Non-Governmental Organization
OECD	Organization for Economic Co-operation and Development
OPEC	Organization of Petroleum Exporting Countries
PRA	Participatory Rural Appraisal
PTA	Parent–Teacher Association
SMC	School Management Committee
UNESCO	United Nations Educational, Social and Cultural Organization
WTO	World Trade Organization

Chapter 1
Introduction

This book assesses the ability of participatory development approaches in aid programmes to reduce poverty and empower disadvantaged communities in marginalized rural and urban slums. To achieve these goals, Third World governments and international organizations, such as the World Bank, have embraced participatory development, which emphasizes creating partnerships and using participatory and people-centred approaches. However, despite the obvious political appeal, the book argues that in practice, interventions of this ilk are yet to have the desired effect. To date, such aid interventions have generally failed because they have tended to ignore questions about inclusiveness, the role of change agents and the personal behaviour of elites that overshadow, or sometimes ignore, questions of legitimacy, justice and power in pursuit of consensus (Kapoor, 2002b).

We advance this and related arguments through an in-depth exploration of a major aid intervention known as the Basic Education Improvement Project (BEIP). The BEIP was implemented by the Government of Kenya (GOK) in partnership with the Organization of Petroleum Exporting Countries (OPEC) and marginalized communities in rural and urban slums to increase access to education, promote participatory development, reduce poverty, and enhance social change and sustainable development. At the heart of the BEIP was investment in educational resources (for example, school buildings) to promote what the GOK refers to as 'balanced development' (GOK, 2003b, 2005a). This was taken to mean " … enhance[d] access and improve[d] ... quality of basic education with a view [to] ensuring the achievement of universal primary education by 2005 and education for all by 2015" (GOK, 2003b, p. 1). The BEIP was a key part of a broader education sector strategy designed "... to give every Kenyan the right to quality education and training no matter [their] socio-economic status" (GOK, 2005a, p. iii). A driving factor underpinning this aim is the view that education contributes to sustainable development by addressing knowledge and information gaps. As described in more detail later in the book, a defining feature of the intervention was that participation was enshrined in policy and practice (or programme design, implementation and monitoring and evaluation).

The aim of the book is to critically examine the impact of the BEIP on disadvantaged[1] people living in urban slums, pockets of poverty in areas of

1 GOK recognize disadvantaged individuals and groups as those who have been historically marginalized politically and by virtue of living in remote areas mostly in

low, medium and high agricultural potential and Arid and Semi-arid Lands (ASALs) in Kenya. In so doing, the value of stakeholders' participation in the BEIP and the conditions that contribute to empowerment and social change are assessed. In advancing this account, reference is made to Ife's (2002) model for community development to assess the 'fit' between policy, practice and theory of participatory development and its relationships with participatory democracy in aid development programmes that engage government partnerships with donors, civil society and local communities.

In the following sections, the book's core arguments are detailed and the background to the study which provides the basis for it is briefly outlined. The chapter concludes by providing an overview of the book's structure.

Participatory Development, Empowerment and Social Change

As described throughout this book, the GOK has embraced participatory development as a strategy to empower disadvantaged communities to take control of their own lives through creating partnerships with donors and local communities. The GOK's focus on participation and partnerships is an indication of political goodwill towards the development of human capacity through approaches to development that emphasize participation. In part, these approaches reflect the GOK's acceptance of arguments advanced in the academic literature (e.g., Murunga, 2002, 2005) that 'bottom-up' approaches are more likely to enable disadvantaged people to direct their own development and to do so in ways that are sustainable. However, while the GOK might profess to advance bottom-up change, little is known about the extent to which mainstream aid programmes in Kenya (or elsewhere) actually adhere to the principles of participatory development. Elsewhere in Africa, only a handful of research (Biggs and Smith, 1998; Botchway, 2001; Choguill, 1996) has evaluated the empowerment and social change value of participation to the disadvantaged in relation to broader issues of structural disadvantage, rights and obligations. For example, in her evaluation of the extent to which the use of participatory methods are at all critical for participatory development to occur, Botchway (2001) found that 'participation' was being used to supplant fundamental structural reforms required for empowerment and social change. Others go further than Botchway, arguing that participatory approaches are in effect tyrannizing development decisions and debates without evidence that they lead to empowerment and social change (Cook and Kothari, 2001).

Unfortunately, studies such as Botchway's (2001) are relatively rare. As a result, in general there is insufficient understanding about what constitutes

adverse climatic, environmental and housing conditions in ASALs, city slums and rural areas. Structural disadvantages based on gender, class (e.g. literate, illiterate), poverty and ethnicity are inferable in these categorizations. Here women, children, illiterate, poor and special needs people are identified as disadvantaged (or deprived).

participation and the conditions that contribute to benefits of empowerment and transformation to the disadvantaged people in sustainable ways. On this problem, Hayward, Simpson and Wood (2004) argue that participatory approaches are being introduced without a clear understanding of how local stakeholders access and experience participation. Similarly, Mikkelsen (2005) recounts that, "of the uses and understandings of participation and associated terms such as 'empowerment', there is no one *apriori* strategy for *who* participates in the development *mainstream*, in *what*, *why* they participate, and *how* and on *which conditions*" (p. 58).

This book argues that a further weakness in the literature is the lack of a clear analytical framework that allows us to properly assess the impact of development. In this respect, Shepherd (1998) contends that participatory development " … has lacked the analytical tools and [an] … adequate theoretical framework" (cited in Hickey and Mohan, 2004b, p. 59). Despite brave assertions that participation is 'political' (White, 1996), the lack of a coherent theoretical framework means that our understanding of participatory development remains insufficiently theorized and poorly defined (Kapoor, 2002b). For example, *participation* has alternately been used to denote participatory methods, participatory development, and even structural reforms that involve collaborative-dialogical processes with stakeholders (Mikkelsen, 2005). Such broad usage is problematic, not simply because it has created confusion, but also because it has led to suspicion that development methodologies, such as Robert Chambers' (1994a) Participatory Rural Appraisal (PRA), have come to replace the actual democratic practice of participation. Indeed, rightly or wrongly, such methods have become perceived as a convenient and controllable substitute for democracy (Biggs and Smith, 1998), and for the kind of change that will actually improve the lives of disadvantaged people.

Nevertheless, in the face of such criticisms, the GOK (2002a, 2004a, 2007) indicates that participation and partnerships in processes of policy-making, programme design, implementation, monitoring and evaluation promote sustainable development and democratic practice. Participation and partnerships contribute to these aims by redefining the roles of the state, private sector, donors, civil societies and subjects of development in policies.

Although the GOK's claims sound plausible in theory, we argue that in practice these approaches are best seen as obscuring from view the fact that poverty (Cornwall, 2002, 2003, 2004) and inequality (Pieterse, 2002) remain major obstacles to realizing the potential offered by participatory development. At a more practical level, the impact of policies designed to increase participation and partnerships and their relationships to implementing mechanisms such as the BEIP has yet to be researched. Considering the scarcity of theoretical and practical information about the empowerment and social change benefits of participation and partnerships to disadvantaged people and the conditions that contribute to such benefits in sustainable ways, the need for research to address these knowledge gaps cannot be overemphasized. Indeed, it is just this kind of information disadvantaged individuals and communities need in order to engage more effectively with the social, political and economic structures that govern their lives.

In a similar refrain, the GOK, donors and civil society and other groups and individuals also need such knowledge to optimize benefits of empowerment through participatory development and participatory democracy to disadvantaged people in ASALs. In his quotation of a Maasai's experience with participatory development approaches in the Kenyan education sector, Nkinyangi (1981, p. 195) states that:

> The fundamental problem of education with pastoral people … is changing their attitude by creating something they believe in. Most pastoral people are not looking for a hand-out; such an attitude is repulsive to them. What they want is something they can really participate in as their own, right from the beginning … it's the whole attitude, the whole approach towards pastoral people that's wrong. People begin with the assumption that these people cannot change. And so they bring in things, sometimes consciously, sometimes unconsciously, that completely antagonize the people and stop them from helping themselves. (cited in Sifuna, 2005b, p. 505)

Central to this problem are issues about how technocrats' decisions implicate the state, donors, and community interests and commitments and how the resulting relationships enhance or limit empowerment and social change to disadvantaged people. One of the challenges with participatory development is that power is frequently delusional and conceals the extent to which participatory processes are manipulative (Cook, 2003) and marginalizing rather than liberating to disadvantaged people. Encapsulating this subtle character of participatory approaches, Tondon (1995) avows:

> … in the name of participation the people are made creators of their own poverty (called development) much like in the way the colonials in Africa used to get people to participate in the building of village roads, and the way people in post-independent Kenya engaged in "harambee" projects, thus relieving the government from carrying out its responsibility to the people. (p. 32)

As noted above, it is clear that change agents can advocate for the 'inclusion' of disadvantaged people in development programmes without necessarily addressing inequalities of class, gender, ethnicity, culture and poverty (Cornwall, 2002). While sensitivity to diversity and inequalities is implied in policies and theories of participatory development, in practice no fundamental changes in the economic, social and political structures that govern disadvantaged people's lives are effected.

In sum, the justification for the book is twofold. Firstly, as noted above, the purpose is to examine critically the extent to which the approaches and principles used in the BEIP contributed to benefits of empowerment, social change and sustainable development to disadvantaged people. Secondly, it assesses the 'fit' between policy, practice and theory of participatory development and their relationship with participatory democracy in aid development programmes that engage government partnerships with local communities, civil society and donors and use this assessment to shed light on theoretical debates that are on-going in development. To assist in this process, the book takes one example of a project,[2] the BEIP, and uses data gathered in interviews, documents, and observations with technocrats, teachers, parents and members of Boards of Governors (BOGs) and School Management Committees (SMCs) to enhance academic debates on practice, policy and theory of participatory development. A total of 60 one-on-one interviews, eight focus groups with a total of 81 participants and participant observations of 557 participants were made in 20 primary and 10 secondary schools. These methods and multiple data sites enhanced reliability of the research design and allowed for thick and rich descriptions of the components of the BEIP by triangulating multiple data sources.

The BEIP was chosen due to its focus on participation, partnerships, empowerment, social change and sustainability. To date, few (if any) studies have explored the impact of the BEIP on disadvantaged people. The BEIP provides a unique opportunity and balanced way to assess the impact of people-centered and participatory approaches within government-donor-led development. The next section considers the theoretical lens applied in this book.

Theoretical Lens of the Book

To explore the issues noted above, the book draws upon Ife's (2002) approach to community development. In his view, there are three ways of achieving empowerment and social change. These are policy and planning, social and political action, and education and consciousness raising. Empowerment through policy and planning is achieved by developing structures and institutions to bring more equitable access to resources or services and provide opportunities to participate in the life of the community. Empowerment through social and political action emphasizes the importance of political struggle and change in increasing power, even in an activist sense of the approach. Here, participation enables people (the hitherto excluded from development) to increase their power through some form of action that equips them to be more effective in the political arena. Furthermore, empowerment through education and consciousness-raising takes into consideration the importance of a broadly-based educative process in equipping people with the necessary knowledge and information. Empowerment incorporates notions of

2 A project is a bounded development plan of action focusing on certain issues of concern to governments, institutions, individuals, groups or sets of special groups. Projects normally are distinguishable by their life span that runs for 3–5 years during which key result areas are expected to manifest.

education and consciousness-raising to help people understand the society and the structures of oppression, giving people the vocabulary and the skills to work towards effective change.

Since empowerment is core to development, Ife (2002) cautions that there are some types of power that need not be sought: power to exploit others, power to wage war, and power to destroy the environment. His view takes into account the fact that the process of participation, as the means of empowerment, can indeed corrupt the (un)anticipated development outcomes.

One advantage of adopting Ife's model is that it offers a holistic and a more balanced way to assess the impact of the BEIP through its ecological and social justice perspectives. Its emphasis on empowerment and change from below provides a unique way to respond to the developmental questions posed by the theories of modernization, dependency, alternative development and post-development.

However, despite these advantages, there have been few, if any, applications of Ife's theory and model to real-life development contexts. Indeed, this book specifically responds to calls for primary research that pays attention to the different contexts and purposes for participation in order to determine what forms of participation are more likely to optimize empowerment and social change benefits to the disadvantaged. It thus enhances debates by paying attention to who actually participates in the development initiatives and who does not, either through exclusion, or self-exclusion (Cornwall, 2000, cited in Mikkelsen, 2005). In this respect, it is widely acknowledged that there is a need to broaden our understanding of how participatory development approaches do or do not utilize indigenous knowledge as a means of building communities on the basis of cooperation to enhance democracy, empowerment, vertical and horizontal accountabilities (Chambers, 2005; Gaventa, 2002, 2004; Gaventa and Valderrama, 2001). We need to understand more clearly the motivations, interests and principles underpinning participatory approaches (as suggested by Hayward et al., 2004). We also need to know more about the ideological origins, typologies and problems with participatory democracy and related approaches in terms of their management and governance (Brown, 2004, cited in Hickey and Mohan, 2004b).

To advance our understanding in these areas, this book first critically describes the objectives and the organizational and management structures used to enact and implement the BEIP. Second, it interrogates the partnerships generated to ascertain their approaches and principles. The focus here is upon whether or not there exists a 'level ground' on which stakeholders participate as equal partners and the extent to which the management structure, participation and partnerships provide an opportunity to empower the disadvantaged without further disenfranchising them. Third, it explores the processes and outcomes of participation with a focus on aims, meanings, challenges and opportunities the BEIP offered towards emancipation and transformation of the disadvantaged. Fourth, it critically examines the extent to which the management structure, participation and partnerships challenged structural disadvantages arising from bureaucracy, culture, gender, age, poverty and broader environmental, economic and political factors that inhibit emancipation and sustainable development.

As stated, the book uses Ife's (2002) approach as a way to understand and appraise broader theories of modernization, dependency and alternative development from which participatory development largely draws. The aim is to valorise the perspectives of disadvantaged people based on their experiences and perceptions. Thus, the book deconstructs government and donor perspectives of participation, empowerment and sustainability and attempts to reconstruct new discourses of the same based on the perspectives of disadvantaged people. The significance of these aims cannot be overemphasized given the current marginalization of the perspectives of disadvantaged people in participatory development policy, theory, practice and academic debates (Ife, 2002).

Background to the Study

The study that lies at the heart of this book was undertaken during a period of relative economic prosperity in Kenya. However, indicators of human development show that economic progress does not necessarily translate into enhanced wellbeing for all, let alone serve in the interests of the disadvantaged people. Since the ascendancy to power of the National Alliance Rainbow Coalition[3] government in 2003, the real GDP growth rate increased from 2.8 per cent in 2003, 4.3 per cent in 2004, 5.8 per cent in 2005 to 6.1 per cent in 2006 (African Peer Review Mechanism (APRM), 2006)[4] and 7.1 per cent in 2007. However, this economic growth has not been matched by improvements in the quality of the lives of disadvantaged people. For example, by June 2007, Kenya's total population was estimated to be 36.9 million, with an alarming 56.8 per cent of the population living below the (relative) international poverty line of one dollar per day (APRM, 2006; GOK, 2005a). These poor people can neither access, nor support, quality livelihoods. In the case of education, for example, the GOK's own evidence is that disadvantaged people have not been able to meet their obligations and actively exercise their right to education, even after the introduction of free and compulsory primary education. The implementation of free and compulsory primary education in 2003 saw a total of 1.2 million out-of-school children enrolled in formal public primary

3 It was a conglomeration of political parties that united to agitate for political change during the 2002 parliamentary elections. Specifically the aim was to oust the then Government of Kenya African National Union which had remained in power for 24 years. When elected, the coalition government aimed to increase democratic space and empower the citizenry with a new constitution that attended to the realities of the diverse Kenyan populace.

4 See US Department of State, http://www.state.gov/r/pa/ei/bgn/2962.htm, accessed on 21 October, 2007.

schools and an additional 300,000 enrolled in non-formal education centres. Yet, 1.5 million children still remained out of school (GOK, 2003d).

Similarly, during the time of the National Alliance Rainbow Coalition government, the adult literacy level in 2007 was estimated at 85.1 per cent, an increase from 82 per cent in 2002 (APRM, 2006, p. 55) and 4.2 million in 1999. Illiteracy represents itself more dramatically among the poor and women (61 per cent women; 39 per cent men). There are also regional disparities in literacy, which are worthy of note. In the Coast and North Eastern Provinces a literacy level of only 37.7 per cent was recorded in 1999 (GOK, 2005b). Other areas of high poverty with a similar trend include urban slums and rural areas, particularly ASALs. These areas experience long spells of drought/famine, diseases and death of livestock (the main source of income). Pockets of poverty in areas of low, medium and high agricultural potential such as among the Luhyia of Western and Luo of Nyanza Provinces experience perennial floods. Lack of infrastructure (access roads, electricity, telephone), insecurity, cattle rustling, poor access to safe water and health care together heighten disadvantages and inhibit sustainable livelihoods in marginalized areas.

Policy documents attest that for the GOK (GOK, 2001b, 2004b) disadvantage increases demand for state support, particularly in education, health and other social services. It also limits economic growth as disadvantaged groups cannot realize their economic potential. As discussed above, the GOK has promoted partnerships between a range of actors as a solution to these and related problems. It is maintained that these partnerships will support sustainable development among disadvantaged people and generally promote participatory development. As detailed in later chapters, the current GOK focus on participation and partnerships can be traced from the poverty reduction strategy paper of 1999 and sector-wide approaches to planning as set out in the Kenya Education Sector Support Programme across the years 2005–2010.

Although the Kenya Education Sector Support Programme is relatively new, participation and partnerships, as components of human development and development cooperation, have a longer history in Kenya. The Sessional Paper No. 10 of 1965 attests that upon assenting to the human rights declaration of 1948 at independence (1963), the GOK entered into a partnership with local communities to combat disease, illiteracy and poverty (structural disadvantages) on a harambee basis. As set out in this policy, the GOK has been struggling to establish an environment conducive to human development. In the case of education, it develops policies, school curricula, employs teachers, and assesses and evaluates the impact of its policies and educational development programmes. Thus, for more than forty years, the GOK has been supporting education as the main route out of poverty, illiteracy-ignorance, and disease, and as a precondition to sustainable development through partnerships of the harambee genre.

In the harambee partnership, communities initiated school development programmes by themselves and only requested government grants to supplement their own efforts. The harambee partnership was founded on principles of a social democracy deriving from top-down approaches to policy planning and indigenous cultural practices of participation. A central theme here was the view that all people participated as equal partners. By bringing their resources together, otherwise disparate groups could work collectively to address issues of human development of interest to the collectivity of society at different levels (family, village, school, district, province and national).

Recent research (e.g., Murunga, 2002) demonstrates that the harambee policy bore political, economic and social meanings. Although rarely acknowledged, harambee was founded on principles of participatory democracy. It was based on the acknowledgement that the state has a responsibility to reduce structural disadvantage and to facilitate access to rights and obligations that were not readily accessible during the colonial era (Saitoti, 2002). In this regard, harambee was meant to facilitate enjoyment of civic and human freedoms (i.e., political independence). To be liberated from ignorance, disease and poverty, communities relentlessly invested time, human and physical capital to enhance educational development. Through harambee, more than 90 per cent of all secondary schools in the 'district and provincial' categories were built (GOK, 2001a).

The impact of harambee was extended through the Gachathi Report[5] (GOK, 1976). Following this report, the GOK intensified direct support to harambee schools (initially built and supported by communities) in terms of policy planning and implementation, curriculum design, and employment and deployment of teachers. Thus, the Gachathi Report not only legitimized harambee schools but also increased government control of them. This legitimation made laudable, unique and distinctive state–community relationship(s) that are rarely acknowledged. According to Chambers (2005), "the gradual legitimation of harambee secondary schools ... and their progressive incorporation into the official education system is an example of a powerful grassroots movement forcing the government's hand, of the tail wagging the dog" (p. 90). This statement represents a glimpse of rich relationships amongst the GOK, donors, civil society and local communities whose research has remained elusive. Research (Murunga, 2002), participatory development practice (Tondon, 1995) and government policy documents (GOK, 2001b, 2003e, 2004b) indicate that over the years harambee has assumed new meanings and has been subject to

5 The Gachathi Report of 1976 documented the findings of the National Committee on Education Objectives and Policies. The report underscored the need to strengthen the education structure and redefined educational objectives and policies giving priority to national unity, economic, social and cultural development. It recommended the setting up of specialized agencies to support and improve the management capacity of the education sector. Arising from this recommendation, the National Centre for Early Childhood Education was established as a constituent part of the Kenya Institute of Education (responsible for curriculum development) and Kenya National Examinations Council (responsible for evaluation of students' achievements). Along with the economic development policy of the time, which was encouraging government support for development initiatives at grassroots, the report also legitimized government support for the harambee schools, which had been initially built and supported by communities.

abuse. A core challenge of attempts to address poverty, illiteracy and disease through harambee is that benefits of empowerment and social change to the disadvantaged people have remained unrealized (Thomas, 1987).

While the harambee interventions resulted in considerable progress, compared to other areas, disadvantaged people in ASALs have not been able to expand educational facilities and to increase participation of their children in education (Mukudi, 1999, 2004; Sifuna, 2005a, 2005b). For this reason, these people have been receiving grants from the government and other donors. Most of the schools have been built and at times repaired by the same donor agencies that built them (Burkey, 1993; Mulenga, 1999). The dependency on government, donor funding, technocrats and elites that resulted has been unfortunate because it has contributed to the belief that disadvantaged people in ASALs are conservative and resistant to change (Sifuna, 2005a, 2005b). Thus they are perceived to be responsible for their own social and economic situation. However, research (Mbaku, 2000a, 2000b, 2004; Nasong'o, 2004) shows that disadvantaged people are not typically the primary cause of the poor conditions they suffer. Murunga (2002), for example, has sheeted much of the poverty experienced by disadvantaged groups back to neoliberalism. In this respect, he has argued that elites' interests in market-based development, both in the public and private sectors, contributed to the implementation of the now 'defunct' cost-sharing policies of the International Monetary Fund (IMF) and World Bank's structural adjustments programmes in the 1980s. As these neoliberal forms of participation gained currency, they arguably negated the 'spirit of harambee' and, through the removal of government subsidies in all social sectors including health and education, cut against broader human development goals. As a result, disadvantaged people could neither afford educational costs and healthcare, nor engage with harambee processes to build schools (as they had previously been doing). Consequently, the cost-sharing policy contributed to increased dropout, and lowered the enrolment and participation rates of children from disadvantaged households in education (GOK, 2001a, 2005b).

This book argues that the convergence in the weakening of participation in harambee and the imposition of cost-sharing policies is significant, not just because it contributed to increased poverty, but also because it opened 'political spaces' in which donors could advance their interests in a wide range of policies, including major public sector reforms. For example, donors have arguably had a defining impact on policies such as those based on the Poverty Reduction Strategy Papers, Participatory Poverty Assessments, Medium Term Expenditure Frameworks of the IMF and World Bank, Sector Wide Approaches to Planning of the Department for International Development (DFID) as outlined in the Kenya Education Sector Support Programme (2005–2010) and Results Based Management of the Canadian International Development Agency as set out in the national strategy for development (or Kenya Vision 2030).

Within these policy reforms, discourses of partnerships, participation, democracy, rights, governance and empowerment have gained new currency. A key impetus driving these reforms comes from the view that if stakeholders participate in policies and programmes that affect them, development will be sustainable. For this reason, the GOK has embraced participatory development and engraved participation in all public sector strategic plans and aid projects such as the BEIP. These participatory strategies seek to enhance achievement of the Millennium Development Goals relating to poverty, education, health and gender. Policy documents (GOK, 1965, 2001a, 2001b, 2004a, 2004b, 2005a, 2005b) attest that these participatory approaches draw upon rights-based perspectives. For example, in a draft strategy, 'Education for sustainable development in Sub-Saharan Africa,' UNESCO (2006) claims:

> … all activities involved should be developed in a holistic approach [where] education activities include concrete actions towards i) poverty reduction; ii) peace and social and political stability; iii) gender equality and equity; iv) health promotion; v) environment sustainability; vi) culture in relation to skills, behaviours and values to be promoted; and vii) the enforcement of the principles of good governance and transparent management. (p. 12)

This holistic approach defines the areas of focus for this particular research and also the different areas of emphasis in the design and implementation of participatory development in the BEIP and its broader strategies. In particular, as described later in the book, the Kenya Education Sector Support Programme (2005–2010) facilitates participation and partnerships through the Sector Wide Approaches to Planning under the auspices of DFID. The Kenya Education Sector Support Programme is a conglomeration of 23 programmes. This strategy holds that participation and partnerships empower the disadvantaged people to control their futures. As part of the infrastructural programme of the Kenya Education Sector Support Programme, the BEIP provided a mechanism for targeting services and addressing the structural and rights deficits of disadvantaged people through participation and partnerships. According to official documents, the BEIP aimed to achieve this by empowering the disadvantaged to take ownership of the policies, decisions and development changes upon which it impacted. The extent to which the BEIP succeeded in achieving these goals will be considered throughout this book. For the moment, a brief outline of the chapters is provided.

Chapter Outline

Chapter 2 reviews the relevant literature, and outlines and critiques key theoretical perspectives of participatory development. The following four chapters present the empirical findings. The empirical aspects of the book draw upon interviews, focus group discussions and participant observations made with members of the Boards of Governors and parent–teacher associations, individual technocrats, teachers and parents who took part in the BEIP. Chapter 3 critically describes the objectives and management structures of the BEIP. Chapter 4 considers the meanings and impacts of partnerships on accountability. Chapter 5 looks into the process and outcomes of participation. Chapter 6 examines the extent to which the management structure, participation and partnerships challenge dominant discourses and structural disadvantages arising from bureaucracy, culture, gender, poverty, environmental, socio-economic and political factors. Chapter 7 concludes the book by summarizing the main findings and their implications for mainstream participatory development practice, theory, policy and future research.

Chapter 2
Theories and Models of Participatory Development and Empowerment

Introduction

This chapter advances the theoretical perspectives drawn upon in this book. It first critically examines contemporary debates about what constitutes development and presents the perspective adopted in subsequent chapters. Next, the strengths and weaknesses of participatory and empowerment models, which donors and governments have used to address the developmental challenges posed by these theories, are critically examined. The chapter draws upon Chambers' PRA (1994a, 1994b), Arnstein's ladder for citizen participation (1969, 1971), Rowland's nomenclature of power (1997, 1998) and Ife's approach to community development (2002). Arising from the book's structural and poststructural perspectives, this chapter argues that we need to (re)politicize development and participation (or participatory development) and to re-imagine empowerment as an open-ended and ongoing process of engagement with longer-term political struggles on a range of spatial scales (Williams, 2004). The conclusion states the advantages of Ife's model to the account offered in this book.

Empowerment and Social Change

Despite a focus on empowerment, social change and sustainability, to date development interventions have not delivered the promised rewards. To address this gap, policy-makers, civil societies, researchers and theorists are increasingly emphasizing the need for democratic practice, participation and partnerships in aid interventions. The argument advanced is that when disadvantaged people participate in policies and programmes that affect them, development is more likely to be sustainable.

As shown below, participatory development has 'blurred' what were once arguably clear divisions between mainstream and alternative development. As a result, a 'political space' has been created for donors, policy-makers, civil societies and disadvantaged communities to participate in what previously would have been considered top-down models of development. This space has allowed these actors an opportunity to engage more effectively with cooperative development practice through advancing discourses of participation, partnerships, democracy, governance, rights, empowerment, sustainability and social change. However, despite the conventional view that these discourses (and related practices) are intrinsically good, contemporary debates reveal that 'development' and 'participation' have assumed different meanings over time and that the concepts remain highly contested. Mainstream participatory development is polarized between protagonists on the one hand and critics on the other. This polarization means that questions of power and control have neither been adequately researched, nor have they been engaged with in practice – despite the need for such analyses being emphasized in the literature (Hickey and Mohan, 2004b, 2005). Where participatory development approaches have been used to empower disadvantaged people, there is neither clear evidence of reduced 'poverty', nor proof that such methods contribute to 'sustainable development' (Cleaver, 1999).

To echo Pieterse (2001), in part, the failure of development to liberate the poor comes from 'theoretical posturing' (or 'pretentiousness') between proponents and critics of mainstream and alternative development about what constitutes development. Such posturing has limited our understanding of the benefits that mainstream and alternative development models have to offer. Indeed, the attempt by proponents to build appeal for alternative development through politics of the 'bad mainstream' and the 'good alternative' has arguably depoliticized development and relegated the significance of inequalities (Pieterse, 2001). It has also arguably concealed the empowerment and social change benefits of participation (Williams, 2004).

Ife's approach to community development offers us a way forward, as its ecological and social justice perspectives provide a more balanced view of empowerment than is currently the case in participatory development discourses. Ife's approach is used to help advance a central argument of the book – that more permanent solutions to structural disadvantages will only be found within reshaped political networks and within structures and interventions that promote the active participation of the government, civil society and the disadvantaged people (Mohan and Stokke, 2000). To put this view into perspective, it is necessary to briefly analyze the meanings of development as promulgated by modernization, dependency and alternative development theories. This analysis will highlight the opportunities and challenges such theories pose to empowerment of disadvantaged people and help us create a theoretical framework for understanding participatory development as experienced within the BEIP.

Development as Modernization

According to Isbister (1991), modernization theories focus on identifying deficiencies in Third World countries, such as the absence of democratic institutions, low levels of economic capital, technology and an absence of modern industries. Speculation is then raised about ways of repairing these deficiencies (Makuwira, 2003). One way of repairing deficiencies identified in the literature draws upon Frank's (1969) 'dual society' thesis (cited in Kapoor, 2002a). Frank contends that developing societies have two economic sectors. One is 'traditional' and the other is 'modern'. Lewis (1964, cited in Willis, 2005) describes the 'traditional' sector as comprising subsistence agriculture and some urban self-employment. The modern sector consists of commercial agriculture, plantations, manufacturing and mining. The former sector is 'underdeveloped' (backward, stagnant and static) because it has lacked exposure to the outside capitalist world, the latter is modern because of such exposure (Frank, 1969). The process of modernization includes the spread of market relations, industrialization through technological diffusion, westernization, nation-building and state formation for post-colonial inheritor states (Pieterse, 2001).

Modernization is, thus, a process through which the underdeveloped nations transform themselves from traditional societies to modern ones through creating economic markets and investing in industry (Isbister, 1991). The development processes entail structural changes in national economies, including a shift away from a rural-agriculture-based to an urban-manufacturing economy (Lewis, 1964), with a corresponding migration of people from rural to urban areas to work in the industries. The associated discourse also encourages the view that metropolitan life is better than rural life. According to Lewis, development takes place when the 'surplus' and un(der)employed people move from the non-profit oriented (or traditional sector) to the modern (or capitalist) sector. The assumption is that 'growth-based' innovation ensures 'mutual benefit' for both the traditional and modern sectors.

Not surprisingly, this view has been widely criticized. Lewis, for example, contends that such mutual benefits cannot be optimized within subsistence economies as these trap people in poverty. Lewis argues further that governments should encourage foreign companies to invest their capital in domestic industrial development through a process of "industrialization by invitation" (cited in Willis, 2005, p. 42). The benefits accruing from the industrial processes of the capitalist sector socialized and integrated the traditional sector in ways that resulted in higher levels of productivity.

Despite presupposing that mutual benefits flow from one sector to the other and that the benefits of development would trickle down to all, capitalism is explicitly identified as a superior vehicle for realizing development objectives. Indeed, in capitalism's latest manifestation, neoliberalism, it is assumed that development is economic progress and the best route to greater economic growth is through redistributive markets and reduction of state intervention and control – letting the market set prices and wages. Such an approach, it is assumed, ensures the most efficient allocation of resources, thereby optimizing growth rates with concomitant social benefits.

Aid development, as promulgated by trade and economic corporations (IMF, World Bank, OECD, WTO, OPEC and other multi(bi)lateral agencies), is seen to advance neoliberalism (Ife, 2002). However, to help mitigate the negative effects of capitalist and market economies on the lives of the poor, these organizations have actually integrated people-centered and participatory approaches in their policies for international development. For example, the structural adjustment programmes of the IMF in the 1980s in Kenya focused on increasing the participation of citizenry in human development. The focus on cost-sharing led the government to reduce subsidies in education, health, energy and other social sectors. It also led to the rationalization of the civil service and encouraged privatization of some government functions and services such as transport, energy and housing. These policies sought to enhance good governance through deconcentrating (or decentralizing) the responsibility of human development away from the government to the private sector and the citizenry.

This book is pessimistic about the containment of nation states and free markets. One source of doubt, for example, is evidence that developing countries (more specifically Tigers of South East Asia) that have made progressive economic growth with concomitant benefits to human development have had a great deal of state control (Willis, 2005). Neo-liberal planning necessarily draws upon the view that development can only be actualized in a capitalist environment and often involves knowledge transfer from developed to developing nations. "Development planning" (Willis, 2005, pp. 34–5) or Eurocentric technocratic (Escobar, 1995) approaches are based on sharing technical know-how from North-to-South (a single path of progress to development and modernization). Ideally, this process entails embracing the social, cultural and economic systems of developed countries – emulating western ways of thinking and doing so as to achieve growth-based innovation which developed countries see as essential to development in general (Burkey, 1993). As part of the process, governments of developing countries are encouraged to emulate corporate styles of management to increase productivity and efficiencies in service delivery (Ife, 2002).

A core weakness of pro-capitalist approaches to development, according to Ife (2002), is that development managerialism ignores questions of governance, which emphasizes processes with a view to increasing the 'rule of development by the people' not the corporate or market system. 'Managerialism', as a technology of governance, enables aid agencies (e.g., World Bank) to use 'extractionist' methods (Chambers, 2002) (such as cost-sharing within structural adjustment programmes) that appropriate the moral value of participation to support modernization's neoliberal agenda (Cook, 2004) without necessarily attending to global inequalities (Pieterse, 2002) and the causes of these (Cornwall, 2002). Where neoliberalism assumes development managerialism (free markets and containment of states), the process of development places technological progress over human development (Pieterse, 2000).

Modernist approaches to development are also criticized by researchers such as Escobar (1996), who argues that the linear-process-based development and dichotomic thinking (traditional and modern sectors) fails to recognize the range of societies in the South and the potentially unique needs and requirements of the local populations. For example, despite erecting monuments of modernism – vast infrastructures and big dams – development, as modernization, is likely to impinge on government legitimacy and local agency. In sum, modernization theory promotes the view that the state's role in development is to maintain law and order within which the market can operate efficiently (Willis, 2005). The belief is that poverty can be reduced through good policies and practices (Brohman, 1996; Isbister, 1991).

However, although theories of modernization raise welcome questions about poverty, they do not adequately address these. Indeed, practices associated with modernization arguably accentuate inequalities (Pieterse, 2002) and have made the situation worse by relegating the role of culture and diversity in development to second place behind economic growth. To defend this view, it is necessary to consider how dependency theorists have responded to the development challenges of poverty.

Development as Dependency

Dependency theories usually work in concert with structuralism (Kapoor, 2002a) to promote state-donor led development. However, as described below, there is ambivalence in the theory about the character of states and donors and the roles they play in addressing issues of development. For example, though critical of theories of modernization and capitalism, supporters of structuralism were not calling an end to Eurocentrism. According to Willis (2005), they maintained that the process of development in Latin America, for example, would take a different path from the one advocated by Eurocentric theorists. Indeed, they emphasized the importance of structure and historical contexts and advocated for national development strategies – greater state intervention to protect national industries from unnecessary competition from the more efficient and established international firms, establishment of local industries and import-substitution, erecting tariff barriers, and local land reforms – to address the inequalities created by imperialism and capitalism. Thus, supporters of structuralism advocated for a 'capitalist-based' development that was suitable for local contexts arguing that the failures of development were but one process of development.

Other dependency theorists offer a different view. They argue that underdevelopment (expressed as dependency) was generated by the very same historical process which contributed to economic development (Frank, 1969, cited in Kapoor, 2002). According to Frank, economic growth in some rich countries has resulted in the impoverishment of the undeveloped world through internationalization of capitalism, which progressively began to grow in influence and dominated world trade[1] (Isbister, 1991, cited in Makuwira, 2003). Capitalism and imperialism are, thus, the primary causes of underdevelopment. For example, based on his analysis of Chile and Brazil, Frank argues that the imperialism and colonialism that accompanied modernization were founded on appropriation of

1 See also Ife, 2002; Pieterse 2001, Willis, 2005 and Murunga and Nasong'o, 2007, for the Kenyan case.

economic surplus. This process integrated into global capitalism even in the most isolated areas of undeveloped countries and created 'metropoles' from which to manage and direct development in the 'satellites' (Frank, 1969). Rather than facilitate mutual benefits from the countries at the 'centre' to the countries at the 'periphery' of international trade, in practice, the role of the nation-state in this context was to accentuate inequalities through exporting wealth to developed nations. At the same time, 'bourgeois' policy-makers and planners were seen as collaborators with imperialism. Their role was to support a shift in power and resources to the west. Frank argued that a shift towards socialism is required to break the power of capitalism and the ties of dependency it creates. This is to be arrived at through revolutionary "class struggle" including "guerrilla warfare" (pp. 371–2, cited in Kapoor, 2002, pp. 648–9).

Frank's view allows space to promote the empowerment and social change discourse from the perspectives of disadvantaged people. However, there are good reasons to be cautious of Frank's view that socialist revolution will eventually promote equality. One reason for this is that Frank (and others who adopt this position) does not offer any viable ways of promoting change through revolution that would necessarily lead to the desired outcomes. Indeed, delinking from the nation-state, as a way to address the failures of capitalist economies and state-led development, arguably inhibits political empowerment while encouraging violence through class struggle and guerrilla warfare. Moreover, other ways exist to address hegemonic powers of international capitalism and of breaking dependency ties in development. For example, Cardoso and Faletto (1979) use a discourse analysis to examine how social groups and practices reproduce and/or resist imperialism in the post-colonial period. To them, dependency is neither secondary, nor the result of an abstract 'logic' of capitalist accumulation, but of particular 'relatively autonomous' relationships and struggles between social classes and groups at the international as well as the local level. To promote development the state can at times seek alliances with multinational agencies. In other situations, the state can promote alliances with local classes and groups to insulate itself from foreign corporate interests. These also differentiate between dependency during the colonial period and the 'new' dependency characterized by the US multinational corporate power.

This is not the place to resolve longstanding debates within development, nor for that matter, within theories of the state. Nevertheless, it is useful to point out that Frank (1969) and Cardoso and Faletto (1979) have different views about the role of the state in capitalist societies. If one accepts Cardoso and Faletto's (1979) view, these different types of dependency mean that, given the appropriate socio-political alignments, dependent relations can generate economic growth in ways that do not necessarily promote underdevelopment: "in spite of structural 'determination', there is room for alternatives in history" (p. xi). In a related position, Cardoso (1973) suggests an "associated-dependent development" and emphasizes that "it is possible to expect development and dependency" (p. 94, cited in Kapoor, 2002, p. 650). This optimism is premised on the view that although structural determination is burdened with western ideologies and may actually promote dependency, it contributes to some economic progress and has achieved progress in human development in areas such as education, health and employment. Nonetheless, Cardoso and Faletto agree with Frank in suggesting that ties of dependency need to be broken by constructing "paths towards socialism" (p. xxiv, 19) and establishing more autonomous development through regional cooperation (Furtado, 1970).

The intention is to modify and work from within capitalism in the short term, at least. However, the challenge is that, though critical of modernization, these dependentistas' views represent *superior* types of modernization because there is not a single modernity but multiple[2] 'modernities' in colonial inheritor countries (Pieterse, 2001). It is worth noting that there exists a contradiction between endogenous perspectives that espouse delinking from international capitalism, global trade and national development, whilst also adopting the ideals of modernization. While remaining critical of liberal modernity, dependency theory suggests that underdeveloped countries can transform capitalism from being an enemy to a saviour (Isbister, 1991). In practice, this leads to increased state control in areas such as economic and human development planning (Willis, 2005). This is problematic, as it does not sit well with a participatory development framework that seeks to increase control of development by the disadvantaged people. Indeed, were we to take Frank's (1969) view, arguably dependency legitimates difference and denies agency to the Third World.

According to Manzo (1991), a further weakness is that by assuming that agency, empowerment and social change are possible from the confines of the nation-state, dependency "treats the individual nation-state in the Third World as the sovereign subject of development" (p. 6). This equation of the political subject with the nation-state makes dependency overlook how its own nationalist and statist inclinations can also dismiss or suppress diversity, agency and culture. For instance, by looking at culture as a subordinate element in their politics, dependentistas neither examine the politics of and within culture, nor are they aware of the ways in which culture frames their own analysis (Kapoor, 2002a). Yet, (in)dependent relationships also entail agents who "are not only great names ... but also small unimportant folk ... as well as policymakers" (Spivak, 1985, p. 254).

---

2 Critics of dependentistas (Frank, 1969) now argue that underdevelopment is not entirely a consequence of western capitalism. Capitalism is not immanent to imperialism. For example, the Atlantic Ocean slave trade which saw West Africans transported to the North and Latin Americas to work in real estate plantations and the Indian Ocean slave trade promulgated by Arabs introduced East African communities (including Kenya) to different kinds of capitalism, long before Imperialism. Pieterse (2001) has also shown that *eurocentrism* represents a multiplicity (communism, easternization, Japanization) of modernities that do not necessarily originate from Europe. Pieterse also argues that ethnocentric and endogenous efforts to development also assume capitalist approaches. Also see Willis (2005).

The next section further analyzes the meanings of development based on how alternative development responds to the questions on poverty, inequality and culture.

Alternative Development

Insight drawn from protagonists of alternative development show that development is economic progress plus human development (Pieterse, 2001). This is because alternative development retains belief in economic growth and accordingly *redefines* development goals to include human development. These redefinitions stem from three fundamental questions which Seers (1972) directs at development:

> What has been happening to poverty? What has been happening to unemployment? What has been happening to inequality? If all three of these have become less severe, then beyond doubt there has been a period of development for the country concerned. If one or two of these central problems have been growing worse, and especially if all the three have, it would be strange to call the results 'development', even if per capita income had soared. (cited in Makuwira, 2003, p. 18)

Proponents have interpreted Seers' questions in different and sometimes competing ways. Generally, alternative development is concerned about poverty reduction to improve people's wellbeing (Chambers, 1997), reduce global inequalities (Pieterse, 2001), and increase cultural agency, democracy, social justice, and empowerment (Freire, 1970; Friedmann, 1992; Ife, 2002), and self-sufficiency and sustainability of development processes (Burkey, 1993). The assumption is that development is not just about economic progress; it is also about "enhancing individual and collective quality of life" (Simon, 1999, p. 2). Again, development is not only a question of physical facilities (e.g., schools, clinics, roads, and dams), but one that is also primarily of people who are in constant relationships with each other and the social, economic and political systems that govern their lives; what development does to people and the way that development is enacted matters (Burkey, 1993). According to Ife (2002), these relationships entwine the environments, methods, processes and goals of development inseparably.

The aim is to increase individual agency in development of the hitherto excluded by avoiding technical and cultural subjection. Alternative development also aims to limit the deleterious effects of capitalism by focusing on the grassroots development. From this perspective, development must be "a culturally grounded process" where outsiders (donors, researchers or technocrats) can neither formulate objectives nor "define what is development outside their own cultural sphere (Martinussen, 1997, p. 45, cited in Makuwira, 2003, p. 14). The idea of culture radicalizes development by assuming that *development is a process of change that is controlled by the people themselves*, not specified levels of achievement (Ife,

2002; Yamamori, Myers, Bediako, and Reed, 1996). Alternative development, thus, espouses developing and respecting people's cultures, knowledge, skills, institutions, economic, social and political processes as well as the people themselves (Ife, 2002).

Here, alternative development seeks autonomy of the nation-state by breaking ties with international capitalist corporations. In contrast to the earlier perspectives, alternative development accords *agency* to disadvantaged people. The locus of development is not the nation-state (as is the case with dependency theory) but individuals at a grassroots level. The focus on grassroots aims to valorize into development theory, policy and practice (Cornwall, 2003; Spivak, 1985) the voices of marginalized people. Here, development is intimately linked with democracy, rights, social justice and empowerment. Accordingly, alternative development gives emphasis to increasing the participation of all. The International Development Bank (Feeney, 1998), for example, argue that:

> Participation in development is both a way of doing development – a process – and an end in itself. As a process, it is based on the notion that individuals and communities must be involved in decisions and programmes that affect their lives. As an end, participation in development means the empowerment of individuals and communities. It means increased self-reliance and sustainability. (p. 8)

Burkey's (1993) definition adds to this point. Here, participation is

> an educational and empowering process in which people in partnership with each other and those able to assist them, identify problems and needs, mobilize resources, and assume responsibility themselves to plan, manage, control and assess the individual and collective actions that they themselves decide upon. (p. 205)

Consequently, *collective action* is necessary to achieve empowerment and to address problems requiring resources beyond the means of the individual. However, since individuals and developing nations lack sufficient resources, Burkey asserts that collective undertakings require an *organizational structure* that is broadly based and which ensures continuity of action independent of individual leadership. This need for an institutional framework partly explains the adoption of participatory and people-centered development approaches in mainstream development by multi(bi)lateral agencies and the subsequent emergence of mainstream participatory development.

Paradigmatically, Pieterse (2001) argues that mainstream participatory development has blurred the neat divisions between mainstream development and alternative development. This merger of theoretical perspectives is evident in Pieterse's (2001) definition of mainstream development as "everyday development talk in developing countries, international institutions and international development cooperation" (p. 94). Here, the UN agencies, the nation-state and its health and education ministries constitute the institutional framework through which to achieve human development. While the nation-state is the context for economic growth, the World Bank and the IMF provide technical support in terms of economic development planning and macroeconomic stabilization.

The role of civil society is to provide "a flexible position of critique" to the approaches the government and donors use to implement structural reforms and human development (Pieterse, 2001, p. 80) through the discourses of rights and democracy. This is further elaborated in the next sub-section.

Democracy and Rights

Taken as a whole, alternative development aims to democratize development processes and build communities by empowering them to gain control over life-enhancing systems and structures (Ife, 2002). On a personal level, it aims to promote self-determination (participation) (Chambers, 2005). To realize these rights on a sustainable basis, Oxfam (cited in Feeney, 1998) states that:

> Participation is a fundamental right. It is a means of engaging poor people in joint analysis and development of priorities. Its ultimate goal should be to foster the existing capacities of local, poor women and men and to increase their self-reliance in ways that outlast specific projects. The purpose of participation is to give a permanent voice to poor or marginalised people and integrate them into the decision-making structures and processes that shape their lives. (p. 7)

This understanding of participation as a democratic right, method and outcome of development brings to the fore questions about how to democratize development in ways that optimize empowerment and social change benefits to disadvantaged groups that constitute the world's largest populace. Just how to democratize development remains contested between critics and proponents of alternative development and mainstream participatory development. Generally, Ake (1996) says that the democracy that is suitable for Africa must have four attributes:

1. People must have some real decision-making power over and above the formal consent of electoral choice. This requires a powerful legislature, decentralization of power to local democratic formations and considerable emphasis on development of institutions for the aggregation and articulation of interests.
2. A social democracy that focuses on concrete political, social and economic rights. This differs from liberal[3] democracy which emphasizes political

3 An empowering democracy for Africa is one that addresses the political, social and economic entitlements of all and more specifically women, children, youth and the disadvantaged peoples without whose participation there can neither be democracy nor

freedoms and rights alone. Social democracy entails investment in the improvement of health, education and capacity to enable people to meaningfully participate in the life of society.

3. A democracy that emphasizes, in a balanced way, collective and individual rights. It must recognize nationalities, subnationalities, ethnic groups and communities as social formations that express freedom and self-realization. Such social formations should be granted rights to cultural expression, and political and economic participation.
4. A democracy of incorporation – an inclusive politics which engenders inclusive participation and equitable access to state resources. It must ensure special representation in legislatures of mass organizations of the youth, labour movements and women's groups, which are typically marginalized, but without whose active participation there is unlikely to be democracy or development (cited in Murunga and Nasong'o, 2007, p. 6).

This understanding of development, as a democratic practice, is closely linked to the view of participation as a right. Both derive from the view that attainment of people's choices and development can be provided in an atmosphere where the government and the *civil society* play an active role (Brohman, 1996; Friedmann, 1992). Again, political empowerment cannot be attained without the aid of the State, since each government (as a component of the State) has a responsibility to develop its people. According to the World Development Report (1997) (cited in Brett, 2003, p. 23), the government has to ensure good governance and sustainable development by fulfilling minimal, intermediate and activist state functions. Governments must address market failure (providing public goods and services) and improve equity (protect the poor). In the former, the government has to ensure defence, law and order, property rights and macro-economic management. It also has to provide public health, basic education, overcome imperfect information and coordinate private activity (fostering markets and cluster initiatives). In the latter function, any government must adopt anti-poverty programmes and disaster relief strategies, provide social insurance, redistribution of pensions, family allowances, unemployment insurance and asset redistribution. The fulfilment of these obligations demands that the government adopts a model of development cooperation and partnerships as:

> … the government cannot choose whether, but only how best to intervene, and government can work in partnership with markets and civil society to ensure that these public goods are provided. (p. 27)

The realization of the roles of government and civil societies in human development show that alternative development and mainstream development goals are not

development (See Ake, 2000, p. 10, also in 1996b, cited in Murunga and Nasong'o, 2007, pp. 4–5).

always contradictory and competing as the literature on participatory development suggests. Mainstream participatory development, for instance, has ushered in both a social justice perspective (Gandhi, 1964; Ife, 2002) and a rights-based approach to development. From Sen's (1999, p. xii) perspective, alternative development espouses the "removal of various types of 'unfreedoms' that leave people with little choice and little opportunity for exercising their reasoned agency."

The move away from development as 'assistance-charity' to 'claims' and 'demands' (rights) means that transparency and responsibility in addition to accountability must also exist (Gregory, 2007). Gregory argues that there is a very fine line between responsibility and accountability. The former entails 'moral choice' and mostly applies to governance. Despite having sensibilities of 'answerability', the concept of accountability is loosely used in current development policy and management. Critical questions linger around the rights-based perspective in aid development management. Given the socio-embeddeness of rights, there are issues with prioritization. Because rights are universal narratives, whose point of view applies in what circumstances?

From this perspective, alternative development could be seen as a *Third System* comprising citizen politics whose importance become apparent when juxtaposed with the failed development efforts of the government (First System) and economic/market power (merchant or Second System). Here, alternative development is development *from below,* where the community is the primary agent and technocrats and donors participate in disadvantaged people's development, not the other way. The next subsection expounds on the view of development from below.

Development From Below

The actualization of 'organic development' that is people-driven, not state-donor-Non-Governmental Organization (NGO)-led, implies a more radical view of development than is currently the case (Ife, 2002). As argued, an ideal alternative development approach assumes *bottom-up* as opposed to top-down development. Realizing bottom-up change demands a reorientation in development thinking. It means that technocrats, international institutions, and NGOs participate in people's local development, not the other way (Chambers, 1997, 2005; Ife, 2002; Pieterse, 2001). Pretty and Guijt (1992) emphasize that to achieve people-driven development requires reshaping of all practices and thinking associated with development assistance. As a process of change, development "will have to begin with the people who know most about their own livelihood systems. It will have to value and develop their knowledge and skills, and put into their hands the means to achieve self-development" (p. 23, cited in Mikkelsen, 2005, p. 55).

Through its critical advantage, alternative development aims to increase the power of the disadvantaged masses over the structures that govern their lives so that they may be able to assume their right and obligations to drive their own futures. However, the approach is not without difficulties. One problem is that protagonists of alternative development have tended to advocate against the 'bad mainstream' and promote the 'good alternative'. Such an approach tends to alienate the disadvantaged from the structures that govern their lives and it inhibits political empowerment. To address these weaknesses, Friedmann (1992) and Brohman (1996) suggest that civil society and donors should work in *collaboration* with governments to attain political will and increase benefits of empowerment and social change. Collaboration is critical, as it will increase access to resources, networks and information by disadvantaged people. The alternative – a developmental view that constantly opposes state structures – means that the disadvantaged will remain marginalized. The continued posturing between mainstream development and alternative development and the failure of both perspectives to deliver the promise of *development* to the disadvantaged people ushered in post-developmentalism.

Post-development draws on the view that, as it has been enacted to date, development of various types has not delivered on its promises. For example, it has failed to achieve a minimum standard of living for the majority of the poor of the world. Indeed, Escobar (1995), a post-development protagonist, challenges the efficacy of organic–natural development in mainstream development aid programmes and questions the whole idea of alternative development. Escobar posits that the concept of 'development' emerged in rhetoric after World War II as "a response to the problematization of poverty" that occurred during this period (p. 45). In this view, development cannot be seen as a product of natural processes of knowledge creation leading to the discovery of the problems addressed therein. Instead, development emerged out of a discursive process governed by modernization thinking. It was believed that development would occur if capacity of the hitherto excluded in development was increased. This resulted in the construction of the world of 'haves' and 'have nots', the making of the First World (developed) and Third World (undeveloped). Such formations and the associated 'dichotomic' thinking promote domination and encourage disregard of local knowledge as a basis for empowerment and social change of the 'have nots' within economic growth theories, neoliberal markets, dependency-structuralist approaches and the alternative development 'paradigm' itself.

Thus, Escobar rejects both mainstream and alternative development not least because of their immanent character, but also because "development was – and continues to be for the most part – a top-down, ethnocentric, and technocratic approach, which treated people and cultures as abstract concepts, statistical figures to be moved up and down in the charts of progress" (Escobar, 1985, p. 44). In a similar refrain, Latouche (1993, p. 161) claims that:

> … the most dangerous solicitations, the sirens with the most insidious song, are not those of the 'true blue' and 'hard' development, but rather those of what is called 'alternative' development. This term can in effect encompass any hope or ideal that one might wish to project into the harsh realities of existence. The fact that it presents a friendly exterior makes 'alternative' development all the more dangerous.

Since "development has been and still is the westernization of the world" (Latouche, 1993, p. 160, cited in Pieterse, 2000, p. 178), alternative development is just covering up for the failures of modernization and dependency. It is the new continuity with colonial administration (Cook, 2003; Escobar, 1995). According to Esteva (1985), development has been based on "irresponsible experiments" which have collapsed indigenous infrastructure, and cultural and social networks upon which the poor depend for survival; it has created poverty (Tondon, 1995). Thus, development should be rejected *tout court* not because it is still driven by capitalism, which is increasingly global in character, and was conceived within modernization, but because the positions of the 'foreign bad' and 'local good' deny the agency of the Third World. Indeed, such dichotomic thinking negates the extent to which the South also owns development (Pieterse, 2000).

However, the failure of development cannot primarily be blamed on Eurocentrism. One reason for this, as Pieterse (2000) argues, is that such a stance ignores the diversity that the term has come to denote. Eurocentrism, North-South and West-East applies to the imposition of external ideologies not necessarily from Europe. Eurocentrism may also refer to undemocratic, managerialistic and paternalistic practices akin to bureaucratic regimes and multilateral donor corporations. For these reasons, the rejection of development in post-development does not mean "an end to the search for new possibilities of change … It should only mean that the binary, the mechanistic, the reductionist, the inhumane and the ultimately self-destructive approach to change is over" (Rahnema, 1997, cited in Makuwira, 2003, p. 19). Post-development, in this case, heralds a new era of inward looking, localization of knowledge and reflexivity, and creates space for grassroots engagement in searching for alternatives to mainstream development practices, which alienate and degrade people's knowledge and culture. In this regard, post-development resonates with alternative development because, as a critical theory, it does not raise criticisms that are peculiar to itself, except for the rejection of development.

The most salient views from all of the debates outlined above, to the arguments advanced in this book (and which protagonists of capitalist economic growth such as Sachs (1992) agree with) is that the 'one-size fits all' approach to development is flawed. Development is also rejected not merely on account of its results, but because of its world-view, mindset and intentions, particularly using developing countries as laboratories of failed development and governance systems (Pieterse, 2001). For example, many governments of developing countries have used western science as an instrument of power, transforming themselves into what critics of modernization discourse call 'laboratory states' (Visvanathan, 1988). So called 'laboratory states', ideally, pave the way for the transference of unresolved conflicts and perceived inadequacies of [the West's] own liberal democratic political systems in the name of participatory development (Kapoor, 2005) into Third World States. This is to say, development has not only failed to provide the anticipated outcomes to developing countries, it has also contributed to an increase in global inequalities and risks that threaten the stability of democratic regimes (Kothari, 1988) and sustainability of global economies in both capitalist and socialist states in the South and North (Pieterse, 2002). This means a more radical understanding of development is necessary if the vision of people-driven or organic development (as opposed to an imposed one) is ever to be realized.

Concomitant to structural and poststructural views, this book supports Escobar's call for alternatives to development (Escobar, 1992). However, to reject 'alternative development' on account that it was 'midwifed' within modernization is to ignore that development practice more often than not precedes policy and theory (Pieterse, 2001). The point is that the terms 'poverty' and 'participation' have a longer history than development itself although rhetoric predates these to post-World War II. In practice, participation in development is universal knowledge, because people all over the world are always engaged in their own development in their own way (Ife, 2002; Tondon, 1995).

This book accepts 'alternatives to development', as a 'pathway' towards the deconstruction of structures and discourses that cause dehumanization and reconstruction of alternative structures and discourses of power and social change from the vantage point of the deprived themselves. This allows 'space' to explore current discourses of empowerment which, ideally, are discourses of domination – views of the deprived are themselves marginalized (Ife, 2002). Starting with Robert Chambers' PRA, the next section explores the models states and donors use to empower the disadvantaged as a way of acquainting readers with the perspectives in this book.

Models of Empowerment

Robert Chambers' PRA

Robert Chambers' PRA framework is important to our understanding of change from below. Chambers (1994b, 1997) argues that to promote the development of disadvantaged people, change agents must transform themselves into learners. They must abandon their top-down attitudes, professional expertise and institutional behaviours. They must constantly reflect on the extent to which their actions inhibit development of their subjects. Chambers assumes that personal changes in the behaviour and attitudes of development practitioners lead to professional changes – taking up participatory methods (e.g., PRA). These will ultimately contribute to institutional change with a culture of information sharing for research and partnerships. From this perspective, participation is a method, process and outcome of development, research and empowerment. Chambers argues that participatory methods are important to get information from the marginalized because most policy-makers are unaware of the needs of the rural poor as most of them live in urban centres and do not share the social circumstances, or social class origins of those they profess to help. Here, development takes place by including those who are previously marginalized in development activities with a view to challenging the biases of development projects that make the disadvantaged invisible (Chambers, 1983). The development process also entails *learning* and *empowering* processes (Chambers, 1994a, 1997) through gaining new capacities and confidence to face realities of social development. Chambers (1983, 1995, 1997) argues that participatory development practitioners must engage reflectively in the process of development. This will allow them to appreciate whether or not they inhibit or promote learning and empowerment. Thus, change agents must engage as *learners* who are sensitive and responsive to local knowledge. Chambers acknowledges that the challenge of change agents is to 'unlearn' their world so as to fit as constructors of disadvantaged people's lives. The development process that is empowering calls for a vision of *transformed relationships* that seek to abolish dichotomies of 'uppers' (technocrats, donors, NGOs) and 'lowers' (disadvantaged people). Thus, Chambers says:

> In an evolving paradigm of development there is a new high ground, a paradigm of people as people … on the new high ground, decentralisation, democracy, diversity and dynamism combine. Multiple local and individual realities are recognized, accepted, enhanced and celebrated. Truth, trust, and diversity link. Baskets of choice replace packages of practices. Doubt, self-critical, self-awareness and acknowledgement of error are valued … For the realities of lowers to count more, and for the new high ground to prevail, it is uppers who have to change. (p. 188, cited in Williams, 2004, p. 560)

According to Chambers (1997), participation empowers the marginalized to challenge the powerful directly. Here participation is not just " … an opportunity to form enduring relationships" (partnerships) but also one that allows us "to confront and transform over-centralized power" (Chambers, 2005, p. 115). Participatory development should also aim to empower "the deprived and the excluded" and enable them to challenge the "exploitative elites" that dominate them through monopolistic political and economic structures (Ghai, 1988, pp. 4–5). Chambers' method of transformation neither aligns with the dichotomic thinking espoused in dependency theory, nor with the socialist endogenous alternatives. Instead, it fits well with mainstream participatory development where government, civil society, donors and local communities engage as equal partners.

There are challenges though to actualization of empowerment. For example, Chambers' view does not clearly define how the anticipated new relationships of mutual empowerment for all involved are to replace hegemonic powers of domination akin to Eurocentricism or Ethnocentricism (Williams, 2004). Indeed, there are winners and losers in current mainstream participatory development practice, despite the use of PRA methods. As such, critics see PRA methods as instrumental-extractive when appropriated in aid development (Cornwall, 2002, 2003) because they neither lead to reduced poverty, nor to sustainable development (Cleaver, 1999). The populist assumption that 'uppers' are capable of changing themselves and that 'lowers' can compete equally with uppers is, thus, paradoxical. It obscures more than it reveals about how uppers are to change. According to Williams (2004), Chambers' view of transformed relationships conceals practitioners' self-interest in the status quo and does not highlight any structural constraints any reform-minded individuals would face in challenging it.

Despite celebration of diversity, democracy and relational dynamism, Chambers' method of transformation is highly individualistic and heavily reliant on voluntarism.[4] Mikkelsen (2005), for example, acknowledges the value of voluntarism, but argues that it fails to realize that in an increasingly globalizing world, not many democratic societies depend purely on voluntary activities to initiate development. As a result, participatory "development practitioners excel in perpetuating the myth that communities are arguably capable of anything, that all that is required is sufficient mobilization (through institutions) and the latent capacities of the community will be unleashed in the interest of development" (Cleaver, 2001, p. 46, cited in Williams, 2004, p. 561).

While appreciating that participatory development tends to treat communities as homogenous and unproblematic in their spatial boundaries (rather than multiple and overlapping), it is helpful to remind ourselves that through self-help harambee projects, communities have contributed immensely to their own educational development in Kenya. What development rhetoric fails to acknowledge is that even though states *pose* as if they are in control and may reflect some form of democratic governance, human development in developing countries, including provision of physical facilities and other infrastructure, has largely remained in the hands of the citizenry. Some of these criticisms against voluntarism, community, local agency on the basis of globalization (either downwards or upwards) are conspiratorial and do not reflect actual experience. To argue that community is often a thing of development projects' making, in which case arbitrary divisions of space are naturalized, and the power effects of these divisions are ignored (Williams, 2004) is not only to deny existence of sub-nationalities and ethnic tribes but also to negate understanding of how culture is reproduced within such boundaries in aid development projects that espouse to empower these groups. As Williams (2004) summarizes, mainstream participatory development stands accused of three major main failings: of emphasizing personal reform over political struggle, of obscuring local power differences by uncritically celebrating the community, and of using a language of emancipation to incorporate marginalized populations of the Global South within an unreconstructed project of capitalist modernization. Contrary to Cook

---

4 Voluntarism denotes the participatory development practice where Community Based Organizations (CBOs) (women or self-help groups) 'choose' to organize around issues of mutual interest without being coerced and contribute resources to achieve their goals. Chambers (2005) argues that in policy planning, there are cases when autocratic decisions should be made to ensure the poor do not miss out in development. This refers to affirmative action and special programmes for disadvantaged people. Voluntarism can, thus, be seen to emphasize choices, self-determination and responsibility towards enhancing living conditions.

and Kothari's (2001) suggestion that the chapter on participatory development should be closed, because it has failed to emancipate the poorest of the global South, these failings suggest that a development approach that seeks to balance global and local perspectives is needed (Mohan and Holland, 2001). Such an approach will arguably repoliticize development and participation (participatory development) by unmasking the repressive structures of culture, gender, class, caste and ethnicity that operate at the micro-scale but are reproduced beyond it (Cornwall, 2002). One of the challenges of achieving this is that participation is defined in micro-economic terms (or redistribution of economic capital), as Arnstein's ladder for citizen participation shows.

Arnstein's Ladder for Citizen Participation

States and donors have used Arnstein's (1969) ladder of eight rungs as a way to understand citizen participation. Each of the eight rungs represents a type of participation and the degree of citizen control over development. Beginning with the lowest two rungs, participation takes the form of manipulation (rung 1) and therapy (rung 2). At these levels, the objective is arguably not to enable people to participate, but to enable power-holders to 'treat' or 'educate' participants. The third and fourth rungs represent participation by informing and consulting respectively. These are levels of "tokenism" that allow have-nots to hear and have a voice but hardly offer the power to ensure that the powerful heed to their voices. There is neither follow-through nor assurance of changing the status quo. The fifth rung is a graduation of participation from tokenism to placation. Placation allows have-nots to advise but the powerful continue to retain the right to decide. The sixth (partnership), seventh (delegated power) and eighth (citizen control) rungs stand for genres of participation that provide citizens with increasing degrees of decision-making power. Citizens can enter into partnerships that enable them to negotiate with the powerful for mutual benefits. Participation at the topmost two rungs enables have-not citizens to make decisions and enjoy full managerial power.

The ladder promotes the idea that participation should allow for:

> the redistribution of power that enables the have-not citizens, presently excluded from the political and economic processes, to be deliberately included in the future. Participation is the means by which [have-not citizens] can induce significant social reform which enables them to share in the benefits of the affluent society. (cited in Hayward et al., 2004, p. 99)

According to Arnstein, participation unlashes the power to achieve individual and collective social development and to advance structural reforms. Ife (2002) concurs with Arnstein in defining empowerment as giving power to individuals or groups, allowing them to take power into their own hands, [and] redistributing power from the haves to the have nots (p. 53). Such definitions call to mind questions about the types of power donors and governments engender to promote through participatory development. Arnstein's view has been instrumental in human development both in the mainstream and alternative versions. In theory, the ladder indicates that there are different degrees of citizen participation (though, in practice, a clear distinction between levels may not be possible). Reading the ladder from bottom to top suggests a hierarchical view that promotes active (or full participation) by all those development interventions directly affect as the goal to be achieved. However, this is a value-laden view that delegitimizes non- or peripheral participation (Hayward et al., 2004). Indeed, the view fails to recognize the value of choice as a form of empowerment in itself.

As can be expected, critics raise objections. Tritter and McCallum (2006), for example, argue that Arnstein places undue emphasis upon the distribution of power and upon the notion of full participation as the goal of development. They point out that concentrating on participation to attain citizen control 'limits effective response to the challenge of involving users. [It also] undermines the potential of the user involvement process [and] ignores the existence of different relevant forms of knowledge and expertise' (p. 156). Although her theoretical emphasis on redistribution of power implies different power types, in practice it emphasizes difference in economic power only. Arguably, Arnstein's view does not take into account the complexity of power and control relations of the process of development, nor does it properly consider how participation in practice actually occurs. Her view fails to recognize that participation is a goal for some users, not just a means. Furthermore, this lack of complexity in the conceptualization of [Arnstein's] model, its failure to consider the process as well as the outcomes, or the importance of methods and feedback systems (Tritter and McCallum, 2006, p. 158) has paid lip service to critical relationships between these and their impact on the anticipated benefits of empowerment to disadvantaged people.

Consistent with a theme of the present book, Tritter and McCallum (2006) argue that there is need for a nuanced model to guide user and public participation in aid programmes that involve government and donor partnerships with disadvantaged people. Such a model should assume that:

> User engagement and empowerment are complex phenomena through which individuals [in a discursive manner] formulate meanings and actions that reflect their desired degree of participation in individuals' and societal decision-making processes; public involvement is likely to fail where there is a mismatch of expectations or method; User involvement requires dynamic structures and processes legitimated by both participants and non-participants. (Ibid., p. 157)

Arguably, these processes are empowering and enabling at the services system, organizational, community and individual levels. They also legitimate the participation of all interested stakes at these same levels. These four levels are particularly important to the book because they indicate levels at which structural disadvantage and social injustices are perpetuated. For example, by assuming that power can be redistributed from 'haves' to 'have-nots', and by attempting to define participation in purely microeconomic terms, in practice, Arnstein's ladder downgrades the role of disadvantaged people in development to that of 'passive' recipients of 'capital power'. As detailed later, such power can in effect be further marginalizing and disempowering to the disadvantaged people.

Despite the merits of this criticism, this book accepts that Arnstein's framework is of value as it advances a sound basis for considering participation as a process of individual empowerment. It also assumes that the 'few' economically powerful can and do make value judgements in favour of disadvantaged people and will be willing to give up power (or change the social order). Such a view provides some scope to address the developmental challenges discussed earlier. Despite these merits, Arnstein's view does not dovetail well with the book's view of citizenship from below. More generally, citizenship from below aims to empower citizens in ways that enable them to claim their participation in development initiatives based on their civic rights. From this perspective, citizenship situates participation in a broader range of socio-political practices or expressions of agency. Accordingly, individuals extend their status and rights as members of a particular political community, thereby increasing their control over socio-economic resources (Hickey and Mohan, 2004a). The combination of citizenship and the rights-based perspective discussed above are likely to enable participatory development to overcome its 'apolitical' and 'localist' nature by allowing citizens 'space' to claim their rights while building on its strengths in the forging of community-based capacity and trust (Hickey, 2002).

Current forms of citizenship and participatory development have not achieved as promised. Arguably, this failure can be attributed to their lack of engagement with broad programmes of capacity building. To date, programmes of capacity building have lacked the ability to empower in a radical way. As detailed later in this book, a targeted civic education of a more political kind than is currently the case is desirable. In this respect, Hickey (2002) notes the problem to date has been the tendency within development to depoliticize issues and strategies of participation, and to overlook the local and historical contexts of citizenship formation in developing countries. He challenges participatory development practitioners, especially multilateral, bilateral and transnational NGOs, to increase their role of nurturing mutual support and social solidarity, or promoting values of social responsibility and reciprocity, of supporting and mobilizing citizenship in the interests of the entire community. The essence of this role is not just partnerships, but it is also that participation requires political and social action to emancipate the marginalized.

A central question about the potential of partnerships to aid participatory development is how the resulting social networks (where these exist) are (de)reconstructed within aid development and what kinds of power they involve. Rowland's classification of power offers some leverage here.

According to Rowlands (1997, 1998) the kind of power we often think about is that used to get other people to do what we want, or the power that other people have to make us do something. This power can take the form of material, markets, education, positions as in bureaucracy etc. This is 'power over'.

It is typically regarded as the most important form of power because it is associated with processes of marginalization and exclusion through which groups and individuals are portrayed as powerless. Other identifiable dimensions of power which should be considered as part of the development process include: 'power to', 'power with' and 'power within.' In practice, there exist very fine distinctions between these dimensions.

Nonetheless, recognition of the diversity of power beyond 'power over' helps to analyze how the policies and strategies used in the BEIP sought to enhance what forms of power to disadvantaged people. It aids the book to balance the way modernization, dependency and alternative development ideals are implicated in the BEIP. The book's view is that, while NGOs, donors and governments may be able to provide a context within which a process of empowerment is possible, it is only individuals who can choose to take those opportunities and to use them. To date, NGOs have tended to use such approaches to accentuate government failures and to accord credit to themselves as being more able to engage with grassroots development than governments. Although NGOs can easily engage with the disadvantaged groups and individuals, their activities are not void of power. Indeed, to say that NGOs are better able to empower, though logical, actually encourages the view that the disadvantaged can depend more on aid than on their own democratically elected government structures for their development. On the basis of participation as a democratic practice and right, donors, governments and civil societies have a responsibility to set up conditions for disadvantaged individuals and groups to empower themselves.

The next section builds on this point by exploring how Ife's approach to community development delineates the roles of the government, civil society (including donors) and communities while highlighting its strengths and weakness.

Ife's Model for Community Development

Ife (2002) provides a more expanded view of development by describing participation as democracy[5] (not as development) to emphasize the political nature of development. His approach is based on the structural and poststructural view that to achieve empowerment, development interventions should engage with the structures and systems that govern people, their development process, and

5 Ife describes democracy as rule of the people to draw the distinction between governance and management.

outcomes. Otherwise, any reforms at the individual and institutional levels will attain limited social change.

This distinction is important. As detailed in later chapters, participatory development encourages managerialism and corporatism rather than governance (or rule of development by the people). Managerialism and corporatism are thus paradoxical. As is the case in dependency theory, these features can advocate diversity, without necessarily affecting change in power (Ife, 2002). An empowering process is one that aims to increase the power of the disadvantaged over life-enhancing structures (Table 2.1), with a focus on the 'conduct' of development.

Table 2.1 Power over life enhancing structures

1.	Power over personal choices and life chances
2.	Power over the definition of need
3.	Power over ideas
4.	Power over institutions
5.	Power over resources
6.	Power over economic activity
7.	Power over reproduction

Source: Summarized from Ife (2002), pp. 57–9.

Ife's view is radical compared to contemporary notions of empowerment as expressed in aid programmes. It takes into account questions of imposition and hegemony through its focus on obligations (including rights/needs), education, participatory democracy [decentralization] and accountability. Empowerment makes the 'heart of humanity' and involves the deprived being able to access and exercise their humanity to the fullest:

> The proposed beneficiaries of development must be active participants in all aspects of the processes that are intended to improve their lives as well as those intended to transform the contexts and conditions within which they must live, and upon which their well-being depends. (Bopp, 1994, p. 24, cited in Makuwira, 2003, p. 30)

As noted, Ife (2002) identifies three ways of achieving empowerment and social change: policy and planning, social and political action, and education and consciousness raising. Empowerment through policy and planning is achieved by developing structures and institutions to bring more equitable access to resources or services and opportunities to participate in the life of the community. Empowerment through social and political action emphasizes the importance of political struggle and change in increasing power, even in an activist sense of the approach. Here, participation enables people to increase their power through forms of action that equip them to be politically effective. Furthermore, empowerment through education and consciousness-raising takes into consideration the importance of a broad-based educative process in equipping people with the necessary knowledge and information. It incorporates notions of consciousness raising to help people understand the society and the structures of oppression, giving people the vocabulary and the skills to work towards effective change.

Since empowerment is core to development, Ife cautions that there are some types of power that need not be sought: power to exploit others, the power to wage war, or power to destroy the environment. This approach also attempts to account for the fact that the process of participation, as the means of empowerment, can indeed corrupt the (un)anticipated development outcomes. It, thus, imbues such processes with an 'integrity' that requires change agents to balance between technical and moral components of development and to seek social and environmental justice.

To this end, Ife's model is based on maintaining a balance between ecological and social justice perspectives. These promote the idea that a sustainable development approach will necessarily engender a balance between local and global perspectives. An ecological perspective values balance (between social, economic and political systems), harmony (mediation of conflict, consensus building (to promote peace and non-violent solutions to potential conflict) and equilibrium (capacity to incorporate opposing positions e.g., personal and political, male and female, theory and practice, conflicting cultures, local and global etc). The rule of equilibrium emphasizes the importance of the relationship between systems and the need to maintain a balance between them. As shown later in this book, this perspective is essential to our understanding of the relational dynamics that came to play through the BEIP management structure and their impact on participatory development and participatory democracy.

Moreover, adopting an ecological and social justice perspective enables the book to critically engage not just with the structures within the BEIP, but also with the broader cultural, social, economic, environmental and political systems in which it was implemented. Such engagement allows space to evaluate the feasibility of empowerment and social change benefits to disadvantaged people based on the actual conditions, experiences and perceptions. A social justice perspective respects change from below, popular agency, rights-citizenship agendas and integrity in the process and outcomes of development.

Ife's model offers further advantages to the book. Its focus on principles of valuing local culture, knowledge, skills and processes of development not only reinforces the book's view on change from below, organic development and democratic practices, but also has the potential to explicate the contradictions between mainstream development and alternative development. Again, Ife's view that the processes of development and participation involve interpersonal interactions and decision making practices, which are not value free, gives it an added advantage over the previous models. In addition, his idea of empowerment, as intertwined in both the process and outcome of development, challenges the reductionist notions in aid programmes that limit empowerment to capacity building as opposed to engagement with broader forms of political marginality.

Ife's model is nonetheless not free of the criticisms raised against structuralism and alternative development. A core area of concern relates to how his framework maintains balance between the personal and collective interests and the motives of policy-makers, donors and civil society representatives to ensure that the hitherto excluded are not further marginalized from development. Participation is used to support elites' interests in aid programmes without necessarily engaging with the root causes of disadvantages (Cornwall, 2002). Despite its stated aim of promoting ownership and sustainable development, participation has, in the main, become a form of tyranny in which participatory development is advocated, but related practices hardly lead to meaningful benefits to disadvantaged people (Cook and Kothari, 2001).

There are further tensions in respect to how donors and governments reconcile their competing interests to actualize participation in ways that empower disadvantaged people. Ife (2002) points out that an irony with aid development is that it is not only disadvantaged groups and individuals that are further disempowered. Thus, although they are portrayed as villains with donors, in the face of international capitalism and global markets, governments have become just as powerless as individuals. After surrendering most of their power to global corporations that determine directly and indirectly what happens in policies (as opposed to democratically elected structures), governments find that they are not able to influence (let alone effectively implement) the social policies they enact to improve the lot of the people they claim to represent. On the basis of this view, donors' practices in relation to the 'conduct' of development and assertions of consensus in development cooperation are questionable. Governments can actually be in the 'driver's seat', but they pretend not to see whose hands are on the steering wheel (Chambers, 2005). Against this background, it is daunting that donor perspectives and the burgeoning literature provided by pluralists and elites at one and the same time applaud 'anti-colonialist' development practices but barely represent a strong engagement with disadvantages.

Conclusion

> These are times of hard-edged, efficiency-driven, competitive management. To talk of honour today is to risk sounding a bit high-minded or romantic – as if there were ever a time when honour was a saving grace in the corridors of government power. We still need strong and principled … government leaders and civil servants. Addressing structural disadvantages and social injustices in fair and responsive ways is the measure of government power. If that challenge is not met, public cynicism about the legitimacy of democratic structures will deepen. The government will continue to be seen as comprising democratic

> structures that are far removed from the realities of disadvantaged people, inhumane … and uncaring to its people. (Gregory, 2007, p. 5)

This chapter has argued that development, as enacted to date, has not liberated the disadvantaged masses. In part, this is because of theoretical and practical posturing between proponents and critics of capitalist economic growth and human development. Such posturing has arguably alienated the disadvantaged from effectively engaging with the structures that govern their lives, and inhibited political empowerment. There is nonetheless consensus in the various views and critiques above that emancipation can barely thrive outside of people-power (participatory democracy), which assumes popular agency (grassroots) or state-national approaches. Development requires a balance between democratic social relationships and economic capital and is heavily reliant on political will (Friedmann, 1992). The convergence of these sensibilities: the realization that it is not just economic progress that matters in development, but that political and social relationships are equally important and, also, the transcendence of the development discourse beyond technicalities to include questions of moral choice, social justice and freedoms, makes the contribution of this book the more significant.

State-led development in developing countries more often than not succumbs to internationalization of capitalism and market power, while endogenous and grassroots development tends to promise more than it can deliver (Pieterse, 2001). The value of Ife's model to the arguments advanced in this book is in attending to participatory development's theoretical and practical 'disconnects'. The main weaknesses of a framework that focuses on community-based services could be summarized as follows: Government(s) may justify the intent of certain policies to reinforce the status quo or hegemony, which may further marginalize disadvantaged groups. For example, by supporting an agenda of government to reduce public spending and facilitate reduction in the share of government spending on human services, community-based services could be described as covert privatization.

Though these weaknesses have been glossed over here (and will be returned to later in the book), they represent a strong critique of participatory development that promotes the idea of community-based services. Notwithstanding these comments, this book contends that Ife's framework provides a more balanced way to contextualize and promote understanding of participation as a democratic process for increasing access to educational services and as a process for empowering the poor than is the case at present. Ife's model has been used in western communities and has not been readily explored in developing countries to identify any similarities or differences. This is an opportunity for this research to contribute insight on how participatory approaches may enhance/hinder empowerment for marginal groups. Ife's framework is likely to at least overcome some of the weaknesses with current participatory development practices in government-donor assisted programmes in three main ways:

First, the framework is neither a 'blueprint' nor a 'prescription' for development. Rather it allows room for adaptation and analysis of empowerment based on social, political and economic, personal/spiritual, environmental and cultural contexts and type(s) of disadvantage.

Second, the framework is based on both structural and post-structural assumptions that "it is the dismantling of the dominant structures of oppression, and the reconstructing of dominant discourses of power", which must be at the centre of any programme's intent to effect empowerment and progressive social change (Ife, 2002, p. 57) through the ecological and social justice perspectives, which are dependent on each other for effectiveness of empowerment. These perspectives speak to the theories of development earlier discussed. They also put into perspective rights-based citizenship and popular participation approaches to development and integrate them into one model through which to empower and transform the disadvantaged and build communities.

Third, the ecological and social justice perspectives mean that issues of marginality are likely to be better contested. Here, agency for demanding and claiming quality services based on civic and human rights is likely to be robust through participation in policy and planning, social and political action, and education and consciousness raising, which the book premise as not independent. Although they may be conflicting, their interrelatedness enables a complementary role in actualizing a vision for a people-driven development perspective.

Generally, Ife's model promotes "understanding [of] how people organize themselves, what their needs are, how policies will impact on populations and what linkages are required ... are key to the success[ful]" (Harper, 1997, p. 776) emancipation of disadvantaged people. The next chapter examines the BEIP structure and its impact on participatory democracy.

Chapter 3
The New Centralism

Introduction

This chapter explores the structure of the BEIP and describes its relationship to Ife's model of community development. It helps address the central questions posed in this book by describing how the management structures established the policy and practical context in which the BEIP was enacted. The central purpose of this book is to draw upon the perceptions of those directly involved in the intervention to explore how the BEIP impacted on participatory development and participatory democracy. To build upon this case, this chapter draws upon data provided by informants to argue that the structure created by the GOK and the principles that underpinned it had a defining impact upon the way the targeted schools (or communities) experienced the intervention. This chapter also sets the scene for subsequent chapters by describing the context in which the selected communities experienced partnerships, participation, empowerment and sustainability.

Building upon the comments made in the introduction to this book, the section below provides further details and analysis of the nature and aims of the intervention. The following sections draw upon documents and the testimony gathered as part of this study from technocrats, members of SMCs/BOGs/PTAs and individual parents to describe the management structure and its contexts, the principles that underpinned it (partnerships, participation, empowerment and sustainability) and assess the impact that these features were perceived to have on participatory democracy. As noted above, these features are identified here to set the scene for understanding subsequent chapters, where the focus is upon the nature of the change achieved by the intervention.

The overall argument advanced in this chapter is that the BEIP promoted decentralization of responsibilities to marginalized groups and individuals, whilst retaining the authority to make decisions at the centre. The conclusion provides a statement on the relationship between participatory development and participatory democracy.

Nature and Objectives of BEIP

An expected longer-term outcome of the BEIP (as stated in the official literature at least) was to promote balanced development (GOK, 2003b, 2005a). This was taken to mean " … enhance[d] access and improve[d] ... quality of basic education with a view of ensuring the achievement of universal primary education by 2005 and education for all by 2015" (GOK, 2003b, p. 1). It is important to note that this outcome would not be achieved by the BEIP in isolation. Rather, achieving the goals of the intervention requires the support of broader policies and structures. The education sector strategy, for example, in which the BEIP sits, states that "the broad objective is to give every Kenyan the right to quality education and training no matter [their] socio-economic status" (GOK, 2005a, p. iii). An important consideration here is that education can contribute to sustainable development by addressing knowledge and information gaps.

While "education is a fundamental right [and] an important [input] to sustainable development, peace and sustainability … " (GOK, 2001, p. 74), it is clearly not enough in itself. Sustainable development also arguably requires the government to enhance the wellbeing of disadvantaged people by attending to the challenges of poverty:

> [T]he government and other partners recognize that the challenges for sustainable development in Kenya are the eradication of poverty and the achievement of sustained broad based economic growth … poverty eradication is viewed not only as a political necessity and a moral obligation but also as an economic imperative for the country's development and raising people's standard of living. (GOK, 2001, p. 73)

Data gathered as part of this research indicate an acceptance that balancing between education and other social, economic and political factors that affect people's wellbeing was more likely to increase benefits of transformation to disadvantaged people. One senior policy-maker attested:

> Education is considered as one of those achievements which will translate into change. But also you must realise that education alone may not [effect change] if the other co-operant factors, or 'development will' is not moving in other [social, economic and political] areas … It does not benefit so much to have so many people who have gone to school and yet they are not given the chance to participate in production of goods [access to economic activity] and services [basic and human rights] or paying taxes [through wages or purchasing goods/services]. (Parsley)

The idea of promoting balanced development by creating an enabling environment for all citizens to enjoy human rights is reflected in the holistic approach technocrats adopted in the BEIP to improve structures and rights of disadvantaged people:

> particularly in ASALs, urban slums and pockets of poverty … The infrastructure programme when developed and fully implemented in a holistic way ... has the potential to contribute towards the achievement of other goals … this will ensure environmental sustainability … proper water supplies and sanitation … gender sensitivity, health [and] hygiene. (GOK, 2005a, p. 3)

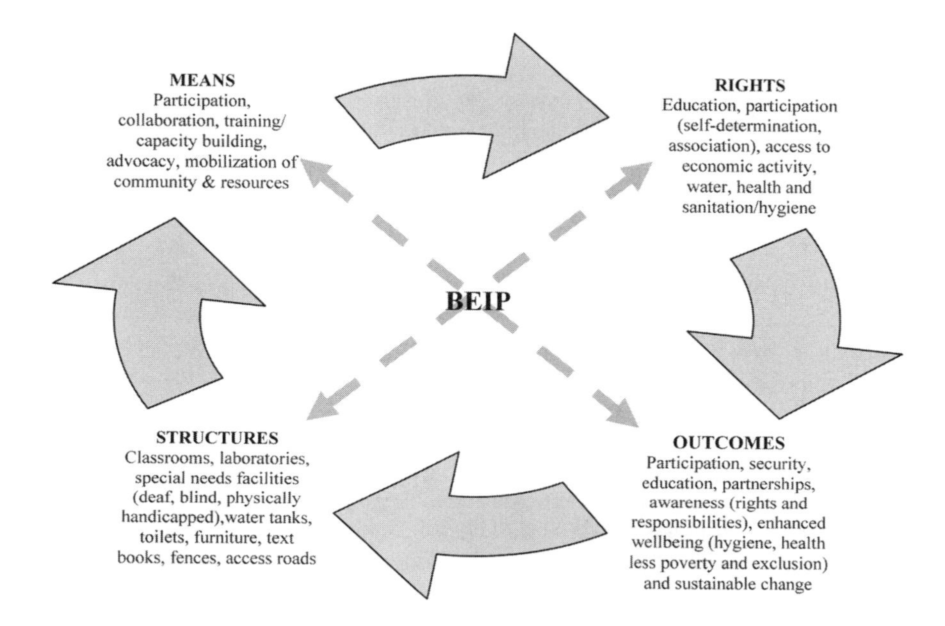

Figure 3.1 The BEIP Structural and Right Perspective of Development

This view that a holistic approach was more likely to increase impact and contribute to achievement of other goals conforms to Ife's (2002) view of holism. Holism underpinned technocrats' belief that the effects of any development intervention spread to the farthest end of the education system. Arising from this belief, technocrats integrated structural and rights perspectives into the BEIP objectives and methods of implementation. Figure 3.1 provides a critical analysis of the interplay amongst the structures, rights, methods, outcomes and process of (holistic and balanced) development that the BEIP was expected to effect on the targeted disadvantaged school communities.

Structures, Rights, Means, Outcomes and Process

Figure 3.1 provides a basis for understanding the intra-interrelationships of the BEIP, with its broader political, social, economic and environmental contexts, in later chapters; the structures, approaches and principles used to support participation, partnerships, empowerment, social change and sustainability; and, the actual outcomes as experienced by those involved. It helps explain how participatory development and participatory democracy were seen to be associated with one another within the BEIP.

The cyclical arrows emphasize the holistic approach and the need for balance between means and outcomes and structures and rights to achieve empowerment and sustainable change. Such balance is manifest in the way technocrats engraved perspectives of structures, rights, methods and outcomes into the BEIP objectives. As reflected in the proposal (GOK, 2002b) and implementation manual (GOK, 2003b), a holistic approach meant using methods that attend to both the structural and rights deficits among the selected disadvantaged school communities:

> The Government of Kenya/OPEC education project will add great value to this programme[1] by increasing access [and] equity and effect quality improvement in basic education through enhancing the built environment of schools and conditions of teaching and learning. At secondary schools level, the focus will be on science education. (GOK, 2003b, p. ii)

In terms of concrete interventions, the project implementation manual (GOK, 2003b) shows that the approach entailed enhancing the infrastructural and teaching and learning environments of 280 primary and 70 secondary schools. Through the BEIP, technocrats aimed to build "1400 classrooms nationwide by 2007" (GOK, 2005a, p. 2) in primary schools. It also aimed to affect the teaching and learning of "science education through provision of science equipment", repairing/building laboratories (GOK, 2003b, p. 1) and enhancing the teaching skills of teachers in physics, chemistry and biology in secondary schools. Another objective targeted the "area … of Special Education, which refers to education and training programmes for children with disabilities. Here facilities and resource provision [was to] be addressed [through] renovat[ing] and well-equipp[ing] Education Resource Centres" (GOK, 2003b, p. ii).

The structures-to-rights process (as illustrated by the double-pointed arrow in Figure 3.1), represents this approach (which is structural in nature, as it treated rights as outcomes). As expressed in the figure, the technocrats claimed to create structures which enhanced the rights of marginalized groups. Put bluntly, the technocrats and their political masters believed that classrooms, boreholes/tanks, toilets and laboratories provided an enabling environment for accessing basic (education, water, health/sanitation and wellbeing) and democratic (or participation) rights. This context, where physical infrastructures form the basis for accessing rights, is closely linked to the rights-based approach (GOK, 2003b), where an increase in access to rights, it was claimed, would ultimately enhance good governance by enacting structures that promote access to physical facilities, more rights, and increase government and community responsibility towards educational development.

This sensibility, as Botchway (2001) and Ife (2002) emphasize, shows that rights and the facilities (or contexts) which enhance these are inseparable. To

1 Free primary education (Kenya's version of universal primary education and education for all).

speak of classrooms, water tanks and toilets in a structural perspective is also to speak of rights to education, water and health/hygiene in a rights perspective. As Ife contends, to maximize its effectiveness, this relationship requires balancing between structures and rights and between methods and outcomes so as to maintain integrity of the development processes that emerge. This point is further illuminated in Figure 3.1 by the means-to-outcomes double-pointed arrow.

As reflected in the broad literature on participatory development (Marsden and Oakley, 1990), the means-to-outcomes double-pointed arrow denotes a process of development where the structures and rights were both methods and expected outcomes. This adds complexity, as means become outcomes, and vice versa. It implies a democratic and empowering process of development. According to Ife (2002), this process is neither linear nor value-free. It entailed decision-making processes which had a normative imperative on the decision-makers and upon the lives of the disadvantaged communities whose rights and structures the BEIP affected.

Through the BEIP, technocrats also sought to increase individual rights to education and, in turn, increase social justice. A strength of this approach is that it conforms to Ife's (2002) contention that the causes and effects of disadvantage extend to the farthest end of a system and so do the decisions, plans and actions of technocrats. In this regard, technocrats' decisions must be informed by the principles of holism and equilibrium.

Nevertheless, despite claims to *holism* (or systemic approach), the practices revealed in the data that are presented in later chapters show significant disconnects with the stated policies and challenges in maintaining a balance between structures, rights, means and outcomes. A core element of this systemic approach arose from technocrats' views of the BEIP, not as an end in itself but as a means of attending to deep-seated challenges of education, poverty, environment, sustainability, water, culture, gender and health in ASALs.

This view that the BEIP provided the context for targeting structures or rights that can be a basis for contesting other rights derives its strength from Ife's (2002) description of participation as a democratic right (or participatory democracy). Technocrats believed that participation was a key democratic right through which disadvantaged communities accessed education, health, economic activity and freedom of association, and thus a core determinant of their emancipation. On the basis of this belief, technocrats first engraved participation into the BEIP objectives and then created structures to enable school communities to collaborate and participate in implementation processes. In this context, participation was also an end and a right in that the BEIP "support[ed] … community participation in the provision of [primary] schools and classrooms, [furniture, pupils' textbooks and teachers' guides] to [enhance] pupils' [academic] performance. [Such participation would ultimately impact students'] health and wellbeing, including support for children with special needs, through providing water and sanitation facilities" (GOK, 2002b, pp. 5–6).

Technocrats also aimed to facilitate outcomes of rights and/or obligations on the part of students, parents and the government since participation in:

> the project [was also seen] to strengthen the management and governance of schools through capacity building and community mobilization. (GOK, 2003b, p. ii)

In this context, participation was expected "to enhance ownership of [decisions, outcomes] and community participation" in the BEIP and future development interventions (GOK, 2003b, p. 2). As evidenced in documents, technocrats believed that collaboration and "participation in the development of education policies and strategies enhances the ownership of the national programmes. This will help us avoid past mistakes" (GOK, 2003b, p. 99). Apparently, these past mistakes entailed implementation of aid interventions by the government, donors and NGOs without due regard for sustainability, collaboration and (active) participation:

> One of the major drawbacks of donor/partner-funded projects [to date] has been [lack] of sustainability. Among contributing factors to lack of sustainability is failure by some agencies and NGOs to incorporate local communities as active participants in such initiatives. This has created the problem of dependence and lack of community ownership of projects. (GOK, 2001, p. 89)

Technocrats' testimonies also emphasized that sustainability was a major concern. Indeed, it was the main reason the GOK chose to implement the BEIP through partnerships and participation. According to empirical data that are presented later in this book, the view is that community participation and partnerships with technocrats and donors would strengthen governance, enhance ownership, empowerment, transformation and sustainability of the changes the BEIP effected on disadvantaged people. Based on this view, technocrats established participation and collaboration as objectives of the BEIP. They also felt obliged to support these objectives with structures (plus user guidelines) that required disadvantaged communities to actively participate and collaborate with technocrats and the donor in implementing the BEIP. Thus, through participation and collaboration, technocrats believed that disadvantaged communities would be empowered to sustain the changes affected through the BEIP and hold educational institutions accountable for delivery of quality services. The strength of this vision is in the belief that participation increases agency and empowers disadvantaged people (Ife, 2002; Williams, 2004). In this case, participation is both a democratic and empowering process and a means by which to increase accountability and responsibility of the government and communities.

The commitment to sustainability is further evident in the way technocrats enacted structures (as elaborated in Chapter 5) to support community sensitization and "mobilization, advocacy, capacity building and training of members of [project coordination units], school [management] committees … boards [of governors] and parents-teachers associations" (GOK, 2002b, p. 5–6). Through these structures, methods and processes, technocrats aimed to empower communities to take greater roles in governance and to increase their participation in society. By increasing participation, in the words of a technocrat, the BEIP was seen to affect social, economic, political and environmental marginalities:

> One of the key areas of focus … was to redress injustices of the past … to increase the participation in education of … marginalized groups particularly in city slums, pockets of poverty and ASALs. (Denise)

In their testimonies, technocrats also observed that by building classrooms, the BEIP promoted the right of education. In the long term, education helped to address other structural injustices (environmental, political, social, and economic) that inhibited disadvantaged communities from fulfilling their education rights and obligations. The attempts by technocrats to address these injustices is reflected in the way they established selection criteria to ensure the BEIP impacted directly on class, poverty, culture and gender. The next section examines the criteria used to select disadvantaged communities, the process of selection and its linkage to the objectives of the BEIP and management structure created to enact it.

Selecting Disadvantaged Communities

As highlighted earlier in this book, the criteria technocrats applied when selecting disadvantaged communities underscores key principles that are important to understanding the impact of the BEIP. These principles are evident in the way technocrats established criteria to select disadvantaged communities where the BEIP would directly impact on poverty. The chapter now looks more closely at these criteria.

Criteria

According to the project implementation manual (GOK, 2003b), technocrats selected schools from areas of high levels of poverty, but which were also heavily populated. Interview data show that the focus on poverty was meant to ensure that the selected schools were actually in dire need of infrastructural facilities (classrooms, water, sanitation and laboratories) to enable the BEIP to increase enrolments (especially for girls) and promote better achievements in education. Lack of these physical facilities and access to the implied basic rights were, thus, indicators of poverty. As noted in the data, educationists believed that schools lacked these physical facilities because the communities[2] that were meant to

2 Official documents (GOK, 2001, 2003, 2005b) acknowledge that through the harambee policy, adopted by the GOK through Sessional Paper No. 10 of 1965, provision

provide them were poor. Thus, by building physical facilities, the BEIP promoted the right to education which also affected other structural disadvantages caused by poverty in the selected schools and neighbourhoods – the BEIP impacted on poverty reduction in the long term.

The need to build the said physical facilities and to enable communities to maintain these facilities helps to explain the close relationship between poverty and the criteria established to facilitate participation, ownership and sustainability. To ensure ownership and sustainability, technocrats selected schools to support on the condition that such schools were "initiated through the action of communities" (GOK, 2003b, p. 15) . This criterion meant that the BEIP mainly targeted schools that had been built by communities (community-based organizations, self-help groups) through the harambee processes described in Chapter 1. Moreover, "community awareness of its needs" (GOK, 2003b, p. 15), as demonstrated in continuous development of the school through active participation prior to being selected, was also a criterion. Other criteria included: "the schools must have potential for expansion to cater for a larger population, must be located within a cluster of schools with a view to spreading the benefits [and affect] gender participation" (GOK, 2003b, pp. 15–16). These criteria resonate with Ife's (2002) idea of change from below and the idea that the disadvantaged communities had significant claims of ownership to the schools even before the BEIP was implemented.

Nonetheless, the findings on the practices that the BEIP generated, as detailed in later chapters, to a great degree represent 'disconnects' between these objectives and Ife's (2002) view of ownership and change from below. For example, the fact that these schools had been built by local communities, arguably, indicate that such communities had (and would continue to take) ownership of the facilities created through the BEIP. However, the policies technocrats enacted appeared to limit community ownership of the decisions and sustainability of the changes upon which the BEIP impacted. This meant there was a lack of harmony between the selection criteria and the policies enacted to promote partnerships, participation, ownership, empowerment and sustainability. It also meant that in some circumstances the outcomes were counter-productive. In addition, local community customs did not always support some key objectives of the BEIP. An instance concerns the education of girls. An educationist confirmed this point when she said:

> There are areas where you want to implement the project but the users do not positively advocate for it because of their cultural aspect. There are areas where education is not taken to be a major thing [for] … girls. That is one of our key focus areas ... Girls should come to school. (Antoinette)

The need to increase the participation of girls in education within cultures that favoured boys arose from technocrats' belief that inequalities based on gender

---

of educational facilities has largely remained in the hands of communities/parents and that expansion of education varied from one community to another.

and culture contribute to poverty. Increasing access to education would stimulate economic growth and significantly enhance equity among ethnic communities. Based on this belief, technocrats established a process which they hoped would ensure equitable distribution of the BEIP benefits by selecting schools from among all ethnic communities and also from the poorest areas. Further detail of this selection process is described in the next sub-section, as it established a disempowering beneficiary-to-benefactor relationship between the GOK/OPEC and the disadvantaged people. This section also sets the scene for understanding the limits and potentials of participatory development methods that espouse microeconomic notions of 'equitable distribution of resources' to empower and socially transform disadvantaged people as detailed in later chapters.

Process

Although technocrats aimed to identify communities that were in the most need and that were most likely to take ownership of the intervention, in practice the selection process did not readily support this objective. The selection process was enacted in two complementary ways (and in two phases), first by drawing upon secondary data (for example, poverty levels), and second through gathering information from representatives. As evident in documents and empirical data, in the first phase technocrats based the selection criteria on the poverty indices of the national census of 1999 (GOK, 2003b). Using the poverty indices, technocrats first ranked districts into categories of high, medium and low income (agricultural) potentials. Next, they identified the poorest of the poor districts in each category. Though logical, the use of poverty indices that largely described national (macro) and districts (meso) levels of economic production in practice concealed environmental, class, cultural and gender differences that are critical to the effective understanding of structural disadvantages at the divisions/locations/villages (micro). Besides, the poverty indices were based only on agricultural potential. This insensitivity to alternative forms of production (and hence, alternative conceptions of poverty) risked accentuating social-economic exclusion rather than addressing it.

As Sifuna (2005a, 2005b) suggests, the data presented in later chapters show that these criteria can advocate for participation and partnerships without affecting socio-economic inequalities. SMCs/BOGs/PTAs perceived these criteria and the process of selection as contributing to dependence on the more agriculturally endowed areas while at a macro level they were seen to entrench perpetual dependence on external aid. This failure to address dependence on aid, a core objective of the BEIP, could partly be blamed on the macro-micro economic principle of redistributing surplus from the more potentially agricultural and the already economically endowed areas to the less potentially agricultural areas on which these criteria were premised.

A related problem arose from the second phase of the selection process where technocrats first relied on secondary data which gave little indication of the actual conditions of the schools as official documents attested:

> Due to the urgency for intervention, the Ministry will not be able to carry out a national survey, but will instead use secondary data in selecting the schools to be assisted. [More] data will be obtained from District Education Officers in target districts who will be required to form district selection committees for the purpose of selecting beneficiary schools. (GOK, 2003b, p. 15)

Technocrats in part relied on secondary data because they were under pressure to submit the project proposal to the donor and to implement the intervention within a timeframe that had been agreed upon by the GOK and the OPEC. Cautious of the misgivings of secondary data, technocrats gathered supplementary information from education representatives in the poorest of the poor districts where the targeted schools were located. These education representatives described the conditions of the schools that they considered to be most needy in their districts as an educationist explained:

> We started by … identifying the schools we are giving [funds] … the district office identified the schools in consultation with the leaders of the local communities … The leaders had to be consulted to avoid this situation where somebody might be saying that this project was taken to the wrong place. This is not right … The leaders are the opinion of the [community] and were fully involved in the identification of the schools. Then after the schools were identified, we did go down to talk to the school management committees to bring them on board and to [let them] know what it is that was expected of them ... parents were also brought on board. (Emmanuel)

While technocrats assumed structural and rights-based approaches to affect structural inequalities, their attempt to use secondary data and representatives to determine the type of intervention, its objectives, and to make decisions about who to involve in what activities and how, had a defining impact on the way the selected disadvantaged groups and individuals experienced the BEIP. As confirmed in the above excerpt, SMCs/BOGs/PTAs neither participated in setting the selection criteria and in the actual selection of the schools nor did technocrats consult the communities directly involved in the BEIP when determining the objectives and selection criteria. This meant that the local communities did not initiate the BEIP, or participate in enacting the policies and decisions that they were meant to implement. Thus, the selection criteria and the related participatory development practices defined SMCs/BOGs/PTAs and the communities they represented as passive recipients of aid, decisions and policies determined elsewhere.

As elaborated below under partnerships, the criteria established fertile grounds for the emergence of a beneficiary-to-benefactor relationship between the school communities and the GOK/OPEC. This relationship cemented bureaucratic power, accentuated market shifts to the donor and further marginalized the very disadvantaged communalities whose livelihoods the BEIP sought to improve. As demonstrated in Chapter 5, in Ife's (2002) terms, the selection criteria and the practices of participation that they generated formed the basis for the separation of processes of planning and policy-making from implementation and monitoring and evaluation activities, which in turn contributed to social exclusion and reduced benefits of ownership to the disadvantaged communities.

Such separation arose partly as a result of the 'haste' that the technocrats were under to implement the intervention and also from technocrats' top-down mindset as evident in the sequence of events. Technocrats first determined objectives, negotiated a loan with OPEC, established the criteria of selection and a management structure, consulted with political/local leaders in identifying the actual schools, and finally informed the SMCs/BOGs/PTAs who in turn informed the broader community of parents and school neighbourhoods. This idea of informing school communities of predetermined decisions so as to gain ownership and support emanated from the hierarchical management structures adopted to enact and implement the BEIP. The next section shows how the management structures, though meant to affect participatory development and participatory democracy through processes of decentralization of decisions and functions, risked legitimating bureaucracy and its undemocratic practices while contributing to a 'new form of centralism'.

Management Structure

This section explores the management structure of the BEIP to highlight its underpinning principles and demonstrate how these features established the policy and practical context of the intervention. According to the project implementation manual (GOK, 2003b, pp. 3–9), the GOK formed a national taskforce and committees (see Figure 3.2) akin to the bureaucratic framework of the Kenyan education sector to enact and implement the BEIP.

As is the case in most, if not all, bureaucracies, the influence of the BEIP management committees stretched downwards from the national office of the Ministry of Education through the district tiers of educational administration to the school (grassroots). Such orientations as top-down (and by implication bottom-up) and 'center-periphery', to use Peet and Hartwick's (1999) term, are critical to our understanding of technocrats' and donor perspectives of mainstream participatory development.

As Ife (2002) contends, these orientations resonate with bureaucracies where the effect is to concentrate power to the centre. Such orientations have also obtained 'capitalist' notions in discourses of cooperation, partnerships, participation and empowerment, where a key aim is to disperse centralized power away from the centre to the grassroots. An important point to note is that bureaucracy is closely

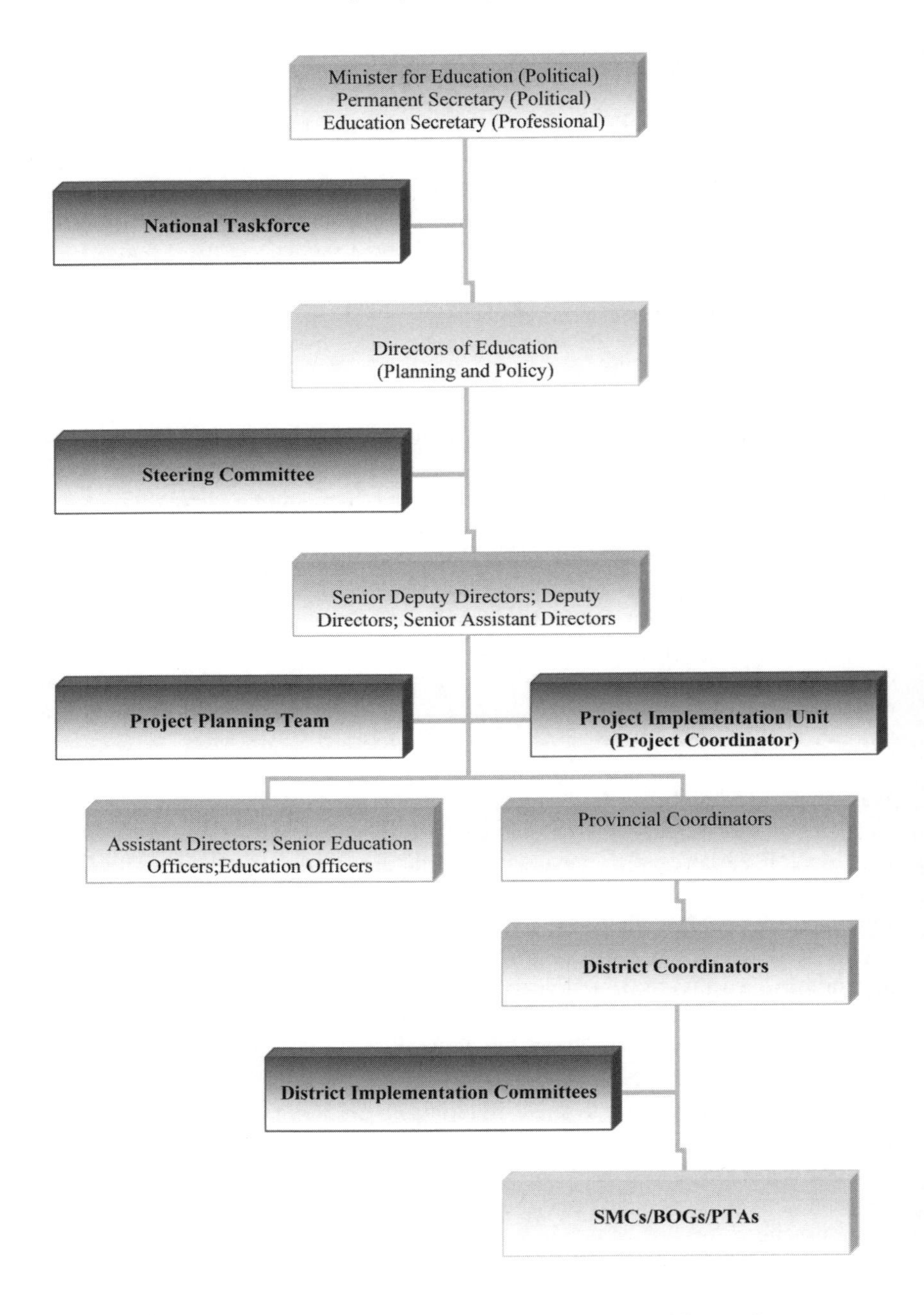

Figure 3.2 Organization Chart of the BEIP (GOK, 2003b, p. 3)

linked with perspectives and practices that seek to cushion individuals and groups from the negative effects of economic change while enabling the government to fulfil its responsibility to the citizenry. In this context, the BEIP management structures presented an expanded notion of bureaucracy. This went beyond the national level (where effect ought to concentrate control to the centre), to include the donor community at the macro level (or supra) and the civil society at the meso and micro levels of development management. This expanded view of bureaucracy is very important to our understanding of how the management structures facilitated mainstream participatory development management within the BEIP.

Official documents (GOK, 2001) indicate that development management through national taskforces and committees is not unique to the BEIP. Again, district education boards, school-based committees, BOGs and PTAs are typical. The Ministry of Education used these to involve teachers, parents, students and broader communities in the implementation of policies and in planning and management of development programmes in schools. As a policy requirement of all development interventions (GOK, 2005a, 2005b), technocrats decided to use these pre-existing structures to avoid duplication of services, functions and resources. Empirical data in this research show that technocrats tailored the formation and functions of the management structures to suit the BEIP objectives, context and nature of the intervention as they perceived them. A key aim in this customization was to affect bureaucracy by generating democratic relationships through participation and collaboration (GOK, 2003b). Participation and partnerships were also meant to support the decentralization of power to grassroots and overcome the bureaucratic and undemocratic tendencies of the education system to which the BEIP structures belonged. Testimonies with technocrats confirmed this aim in the words:

> … when they said the money be taken to the schools through many channels, such as ... the money would come from the donor through the Central Bank to the ministry, to district and to [school], we sat back and said no. Let's cut [down] on bureaucracy and have this money channelled directly to the school. (Antoinette)

This understanding of bureaucracy developing beyond the conventional organizational structure of the Ministry of Education to include the OPEC gave rise to different, but interwoven, types of power in the BEIP. These were: bureaucratic power, which conformed to the organizational framework of the education system; technical expertise, which defined reference authority that comes with professional and knowledge claims; corporate (or aid) power, which emerged through the processes of participation and partnerships; and the 'power(lessness) of poverty', which provided the basis for the former three.

According to technocrats' testimonies, bureaucratic power was entrenched in the management structures. The organization and membership of the management structures drew upon the belief that, by virtue of their positions and authority, senior bureaucrats were capable of effecting feasible relationships of management, collaboration and participation as the words of Antoinette revealed:

> To start with, we involved the key individuals and set up a structure … To ensure that the relationships were workable we got a representative from the very high cadre of individuals at the (re)design level. In every organization that we intended to use in the OPEC [project], we started by having a national team … in [which] we involved the heads of various departments. Once you start by involving the heads and make them meet and interact … the next structure that we set was that of their *juniors*. So if the *seniors* have already accepted to interact with each other as stakeholders, their relationships are okay, then the juniors will follow, and this is what will trickle down to the *lowest* level … That is at schools … you still have the same relationships because it [the structure] is having support from the *top* level management.

The emergence of bureaucratic power was no surprise, as it represented the dominant organizational culture of the education system within which the BEIP management structure operated and where power binaries of 'seniors' and 'juniors', 'topmost' and 'lowest' resonate. The legitimation of bureaucracy was also evident in the organization of the management structures. As shown in Figure 3.2, these management structures emphasized vertical relationships in theory and practice by placing the taskforce (political elites and senior technocrats) at the apex and SMCs/BOGs/PTAs at the base. Indeed, the data show, in practice, authority in the BEIP resided with political and professional elites within the government. In tandem with the need to affect decentralization and empowerment (see Chapter 6), the assumption is that these political elites are capable (and willing) to cede such authority to those lower down the hierarchy.

The belief that power resided with senior bureaucrats not only encouraged the view that the most senior technocrats are most able to enact effective decisions and policies, as they were the most knowledgeable and experienced, but it also led to the emergence of technical expertise power. Technical expertise power is evident in the way the management structures established the context and nature of the intervention. As shown in Figure 3.2 above, the taskforce and the national-district-based committees extended outwards, away from the vertical line manifested in the centralized structure of the education sector. The way the management structure defined horizontal relationships through participation and partnerships contributed to emergence of technical expertise power with the knowledge claims of the participants in the BEIP. This power was manifested in terms such as 'technical experts', 'literate' and 'illiterate.' It determined who to include or not include and defined the roles the 'invited' participants played based on their knowledge claims. As detailed in Chapter 4, technical expertise power was premised on the belief that the poor did not understand what was in their best interests and that their knowledge gaps needed to be bridged by technocrats. Arising from this view, technical expertise power was closely linked to professional/technical competencies of architectural designers, quantity surveyors and other infrastructural engineers who played an active role in the BEIP. As such, the knowledge claims of the disadvantaged people were marginalized in the management structure despite the BEIP objective of increasing their role in the management and governance of their schools.

Corporate(ism) power also aided marginalization. Corporate power emerged with the need for aid assistance and was a critical factor in defining allocation of resources to the selected communities. As such, it represented not just material power (or surplus), as illustrated in the use of terms such as 'distribution', but also through 'collaboration' and 'sharing' of ideas, corporatism power worked in concert with the previous two powers to apportion responsibilities to technocrats, donor and the disadvantaged people.

While the workings of these forms of power are elaborated in subsequent chapters, suffice to state here that there is a very fine line between these types of power as they all represented aspects of domination, inclusion or exclusion. Bureaucratic, technical expertise and corporate powers conform to Rowlands' (1997, 1998) dimension of 'power over'. In the case of the BEIP, they proved problematic because they stifled grassroots involvement. These powers were partly responsible for the exclusion of disadvantaged people from decision-making processes. Though meant to support the BEIP objectives of participation, partnerships, empowerment and social change, technocrats appropriated these types of power to dominate those perceived to be poor (or powerless), maintain balance, and reinforce the status quo.

Overall, the way technocrats appropriated these powers determined whether, through the practices of participation and partnership, the disadvantaged communities would reveal their powerlessness or manifest 'power to' see possibilities of change, 'power with' which partnerships placed on an equal basis would emerge and 'power within' themselves to see and build their self-worth and self-esteem. These powers, according to Rowlands (1997, 1998), epitomize inclusion and degrees of emancipation and sustainable change. As detailed in subsequent chapters, the problem is that the management structure barely presented the disadvantaged people with concrete opportunities to demonstrate, in Ife's (2002) terms, 'power over' personal choices, definition of need, ideas, institutions, resources, economic activity and reproduction. Instead, the management structure legitimized bureaucracy through the way technocrats defined who should participate and who to collaborate with and the roles they assigned to the selected groups and individuals. Rather than challenge bureaucracy, such practices reinforced it and its undemocratic nature. Starting with the national taskforce (Figure 3.2), the next subsection explores the formation and specific functions of these management structures. It argues that by retaining decision-making authority with the topmost bureaucrats, technocrats created a context in which centralism could flourish.

National Taskforce and the New Centralism

The national taskforce is the topmost management structure in the BEIP. According to documentary evidence members of the taskforce were drawn from the Ministry of Education and other sectors of government whose core functions relate to the project objectives. These included permanent secretaries of the Ministries of Education (chairmen), Health, Finance, Planning, Public Works and Housing (political elites). Other members from the Ministry of Education included the chief economist and chief finance officer, principal procurement officer, principal accounts controller, senior deputy director (planning and policy-formulation) and the project coordinator.

The composition of the national taskforce epitomized the view noted earlier that the most senior bureaucrats are most experienced, knowledgeable and better able to effect workable networks of change for the selected disadvantaged people. Besides accentuating bureaucracy, the composition of the taskforce shows that it was the highest authority in decision-making because its members represented such authority in their respective areas of specialization. As part of this authority, the taskforce made the most important decisions relating to policy, resources, planning and implementation of the BEIP. It approved work plans, budgets, disbursement of funds, monitored progress, and advised on policy changes in the course of implementation.

The vesting of decision-making authority in the taskforce had a defining impact on the way the disadvantaged communities experienced the BEIP. It meant, for example, that the management structure facilitated devolution of responsibilities (functions and services) from the taskforce to the structures lower down the hierarchy through partnerships and participation. As argued later in this chapter, such 'decentralization' of responsibilities without a concomitant authority to make decisions created a *new centralism.* How the new centralism was promulgated is described below. For the moment, attention turns to the Steering Committee to which the National Taskforce delegated such decision-making authority.

Steering Committee

The Steering Committee mainly consisted of education professionals. These were the directors of education (including the director of quality assurance and standards) and senior deputy directors. Other members included representatives from the Ministry of Education, semi-autonomous government agencies[3] and bilateral organizations.[4] The Steering Committee guided in planning and it provided policy advice. It also coordinated training (or capacity-building) of the

3 Kenya Education Staff Institute, Kenya Institute for Special Education, Teachers Service Commission and Kenya Institute of Education.

4 Kenya National Commission for UNESCO, UNESCO regional office.

members of the SMCs/BOGs/PTAs, as well as the project planning team and those in the project implementation units. The Steering Committee also advised on the selection of schools, and the establishment of reporting and community mobilization systems.

The Steering Committee established collaboration networks between educationists, technical experts, planners and OPEC to enact the BEIP. It also negotiated the BEIP loan/fund with OPEC using delegated decision-making authority from the national taskforce. To enact its guidance roles, the Steering Committee mainly delegated functions and services to the project coordination unit. It, however, approved how the coordination unit enacted such services and functions. That means the Steering Committee made the final decisions. In practice then, there was a very fine line between the Steering Committee and the national taskforce, which delegated decisions, functions, defined how the selected communities participated, and how collaboration with other sectors was facilitated through the project planning team.

Project Planning Team

Similar to the national taskforce, members of the planning team represented core departments upon which the project aimed to have an impact. These were departments within the earlier said semi autonomous government agencies and those in planning and policy formulation, quality assurance and standards, primary, secondary and special education within the education sector. The role of these representatives was no different to that of the Steering Committee except that members of the project planning team were middle level managers who had no authority to make decisions. Again, members were not permanently attached to the project. Instead, the project implementation unit liaised with the particular departments when their services were needed. By drawing members from these departments, technocrats aimed to increase responsiveness of the BEIP to educational needs as construed in the represented departments and impact on partnerships and social networks across departments.

As part of its roles, with guidance of the taskforce and Steering Committee, the planning team developed the project implementation manual which defined BEIP management structures, memberships, reporting and evaluation systems. It also developed the modules used to train SMCs/BOGs/PTAs on their roles, established the criteria which the national taskforce approved for use in the selection of the disadvantaged schools, and monitored and evaluated project activities as delegated by the structures higher up the hierarchy. Arguably, these roles defined the planning team as an operational management structure in matters of coordination, facilitation, training, monitoring, evaluation and networking. As detailed later, the planning team could be seen to heighten bureaucracy and entrench the dominant organization culture particularly because its roles were also performed by the project implementation unit.

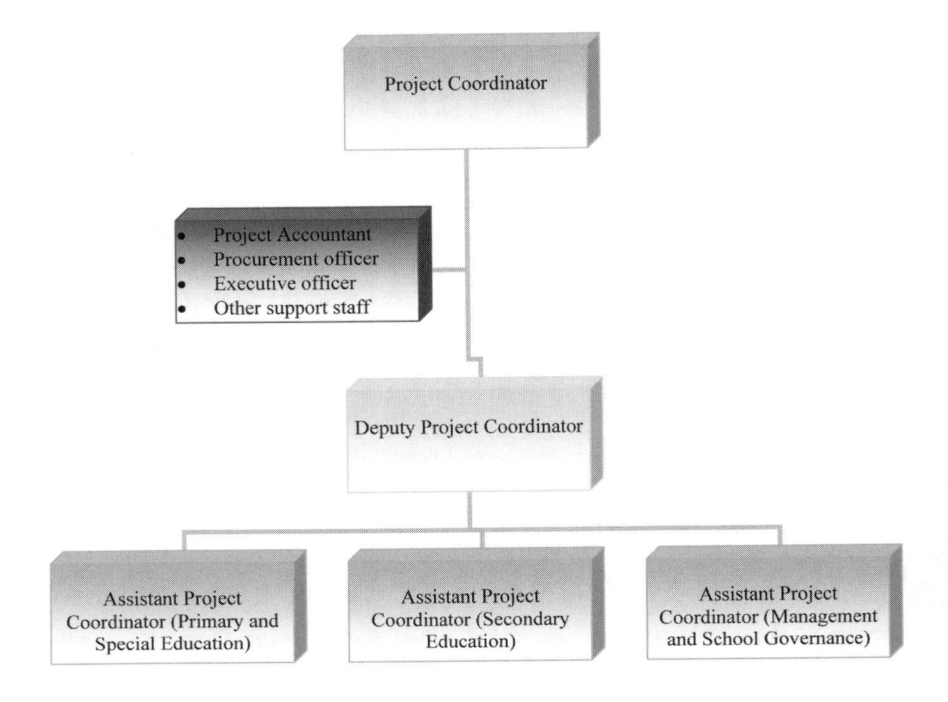

Figure 3.3 Management Structure of the Project Implementation Unit (GOK, 2003b, p. 9)

Project Implementation Unit

The roles of the project implementation unit were to administer and coordinate BEIP activities. As shown in Figure 3.3, it consisted of a project coordinator, a deputy project coordinator and three assistant coordinators. The project implementation unit also had a procurement, accountant and executive officers who were seconded from their respective departments to perform such duties as the need arose. The functions of the assistant project coordinators were closely tied with the BEIP three main outcome areas: increasing access to primary and special education, secondary education, and enhancing good governance in educational systems at the school, district and national tiers of development management. As part of the project implementation unit coordination role, the project coordinator was also a member of the national taskforce, Steering Committee and convenor of the planning team.

The extent that the project implementation unit coordinated but played a cursory role in decision-making helps to further explain potentials and pitfalls of participatory development management through hierarchically structured committees. In practice, the project implementation unit convened all meetings involving implementation of the BEIP. Arguably then, and in bureaucratic terms, the project implementation unit was the most important structure which ought to possess the decision-making authority, not the taskforce and the Steering Committee. As explained later in this chapter, while in theory these topmost committees possessed decision-making authority, in practice, they only 'approved' plans and decisions as enacted by the project implementation unit.

District Implementation Committees

District implementation committees derived their mandate from the legal framework of district education boards. Since their establishment in the 1950s, district education boards have over the years increased their role in overseeing educational development within districts. While the project used the services of district boards, technocrats established parallel committees to oversee BEIP implementation in the districts. Committee members included the District Education Officer (Chairman), District Inspector of Schools (Secretary) and the most senior officials in areas of procurement, social development, public works, health, water engineering and school accounts. These advised, monitored and evaluated BEIP progress in the schools (GOK, 2003b). Technocrats' testimonies in this research also support these roles:

> We [project implementation unit] … first formed the district technical teams. These involved the public works officer, the water engineer [and] coordinator of the project. The idea was that these people are actually acting as mediators. The District Education Officer has a coordination role. So these are people who are given money to visit, like if you [SMC/BOG/PTA] are going to sink a borehole or build a water tank, you ask the coordinator, the water engineer and public works person to visit the area. In addition, they also take the health person. These are the people who kind of supervise the project on our [National-based structures] behalf. You know the parents are there [at school] on a daily basis … but [district committees] are people who come once a week or once in two weeks. So they actually come to see how is the work progressing? And they are able to come back to the headquarters if there is a hitch … If the parents are experiencing problems or are lacking technical know-how, they are the ones who advised them, 'this is the way you need to move.' So they have been very useful. I remember a case where a public works officer was [drawn] from the province because they could not get one from the district. They wanted … advice about how to go about it … we advised [them to] get a works officer from the province. (Carla)

In addition to coordinating BEIP progress, district implementation committees also played a liaison role between SMCs/BOGs/PTAs and the project implementation unit. These also provided technical expertise to the SMCs/BOGs/PTAs on matters relating to construction of classrooms, water tanks, sanitation facilities and other infrastructures. As they undertook these roles, the district coordination units were able to collaborate with SMCs/BOGs/PTAs. In an attempt to ensure the creation of new social networks in which disadvantaged groups participated, the GOK drew representatives from the sectors of education, water, health and housing. Although the core functions of these sectors had a critical bearing on the BEIP objectives, as detailed in later chapters, this arrangement established a fertile ground for elite-to-elite networks in which the disadvantaged people were excluded and their knowledge devalued.

Boards of Governors

BOGs are established through the Education Act to assist in the management of secondary schools and primary and diploma teachers' colleges. Members are appointed by the minister for education. Membership is changed every two years. Key stakeholders that must be represented in the boards include PTAs and school sponsor(s). Sponsors are typically religious or community-based organizations (GOK, 2005b). Board members are normally drawn from the local communities that build the schools, members of parliament, business people and old-boys' and girls' associations. Initially, schools drew board members from close proximity. However, as secondary schools became multicultural, their management also became complex. This has come to mean that BOG members reside in rural and urban areas where they earn a living. This rural and urban dichotomy has cost implications to the BEIP. As detailed in a later chapter, it is also an essential element through which to understand how the BEIP affected knowledge of disadvantaged communities and inequalities of class, culture, gender and environment.

Meanwhile, the GOK did not establish parallel boards in the target schools. It enacted a policy that required serving board members (chairman, head teacher (secretary), treasurer and chairman of PTAs to automatically assume management roles in the BEIP. To further tailor the BOGs to the objectives of the BEIP, the GOK also ascribed additional members who were not necessarily board members – three science teachers, area chief/councillor and community/opinion leader. Such customization together with the legitimation of the BOG by the minister as detailed in Chapter 4 encouraged upward, not downward, accountability. Their management roles entailed identifying needs, implementing construction plans, managing project finances and contracts, monitoring progress, reporting and accounting for expenditure, and sensitizing the communities (GOK, 2003b).

Parent–Teacher Associations

PTAs are the second arm in management of secondary schools. Like BOGs, PTAs are not unique to the project. Documents describe PTAs in terms of "community representative organ with no ties to the ministry of education head office" (GOK, 2001, p. 65). PTAs were initially established in the 1970s through a presidential proclamation. Their establishment is based on the view that secondary schools need to have a body representing the voice of parents and teachers outside the Ministry of Education and the BOGs. Again, parents and teachers interact with students and the community more closely than technocrats and board members. Indeed, parents and the community are the real owners of the schools. According to the harambee policy (GOK, 1965), initially, the role of PTAs was to raise funds for the construction of harambee (secondary) schools. However, because they lacked a legal basis, PTAs could not manage the funds they raised. Instead, the BOGs assumed this role. This created a conflict of interest, which led PTAs to demand to be represented in the BOGs so as to be able to monitor expenditure of their funds.

These demands have led the GOK to ratify the institution and role of PTAs in school management and planning through the Sessional Paper No. 1 of 2005 (GOK, 2005b). This policy has reinforced the participation of PTAs in BOGs, albeit through representatives. PTA representatives provide checks and balances on the way BOGs implement school development plans and manage financial resources on behalf of parents. To avoid conflict of interests, the BEIP technocrats used the BOGs in secondary school. To ensure parents'/teachers' interests were valorised in the BEIP, the GOK prescribed representational positions for the sponsor, PTA chairman and science teachers in BOGs, as Carla confirmed:

> We … have a PTA chairman [who is] representing the whole PTA. These people we keep telling them 'go back among your people. Sell the idea about what is happening. It is not the project of OPEC … it is a project of the school. Let them [community] know so that they can own it … [Although we have] the BOGs [who are appointed by the minister], we go outside that body and take members of the PTA who are now the parents. So it is like all of them have been brought together. You have harmonized their efforts. BOGs are for management and PTAs are representatives of the parents. [PTAs] keep going back to their communities. When they have … parents and sports days they tell their friends the new developments in their schools.

This distinction that PTAs are elected by the communities they represent, while BOGs are appointed by the Minister, has direct implications on the way the GOK defined the composition and functions of the management structures to satisfy the BEIP objectives. The point is that, ideally, PTAs were intended to represent the interests of wider communities of parents and school neighbourhoods. In practice, they were considered to be more legitimate compared to BOGs, not only because communities/parents elect them but also because the act of electing enhances legitimacy and responsibility of representatives. That is to say, PTAs provided parents with a forum to exercise 'voice' and 'choice' through representative democracy and participatory democracy as shown in the formation and functions of SMCs.

School Management Committees

According to the GOK (2005b), SMCs were established to oversee the management of primary schools. Unlike BOGs, they are relatively free to run their institutions with little reference to the National Education Office. More often than not, they work with the Teachers Service Commission and the District Education Office on matters of deployment of teachers and school development respectively. The BEIP operated within an existing legal framework of SMCs (GOK, 2005b). As is the case of BOGs and PTAs, not all members assumed management roles in the project. According to the terms of reference (GOK, 2003b), the BEIP targeted offices of the chairman, treasurer, head teacher, senior teacher, area education officer and area chief/councillor/village elder. Their roles show in the data. Carla said:

> We have found SMCs very useful in financial management … these make sure that the money is used properly. They collaborate with the head teacher to see how the money is being used, what materials are being bought. In addition they also supervise. They come around to see how the structure [building] is coming up … OPEC is a project which is involved in construction of classrooms, sanitation and water facilities. So we have found these parents very useful in coming to supervise, making sure that what has been bought is being used properly and what they agreed in meetings is what is being implemented.

These roles are similar to those of the BOGs. Unlike BOGs, SMC members are parents of children currently enrolled in their respective schools. Parents typically elect SMCs from a group of guardians/parents who oversee the interests of students belonging to the same class (year of study). Carla attested to the process of electing SMCs and the way it legitimates representational roles by saying:

> These are really representatives of the community … in the SMC … what we have done is … in every primary school … every class has representative parents. These end up forming the school committee. So if you are a parent representative of say standard one children, the chances that he is representing the wider group are very high because he was elected by the parents of that class and he is duty bound to report back to them on new development in his class or school as a whole. I found that very useful because you are telling a representative of that class … so these parents came together to elect their representative, so he is truly their representative …

The belief that the act of electing enhances legitimacy and responsibility of representatives is critical to our understanding of how the management structures facilitated democracy and its impact on partnerships, participation and empowerment. Having described the structure of the BEIP, the discussion now shifts to its impact on participatory democracy.

The BEIP and Participatory Democracy

As noted, the way technocrats established the management structures and defined roles of representatives attest that the national-based, district-based and grassroots structures (SMCs/BOGs/PTAs) were meant to facilitate democratic processes, increase benefits of ownership, harmonization of functions and services, and raise awareness of rights through advocacy, participation and partnerships. The project implementation manual (GOK, 2003b) shows that technocrats created committees at all tiers of administration to enable disadvantaged communities to increase their role in management and governance. They also created structures and mechanisms designed to facilitate participation and collaboration. Ultimately, it was hoped that these features would promote "transparency, decentralization, teamwork and performance-based management and accountability" (GOK, 2005a, p. xii). In doing so, the management structures should be seen to establish the context through which to promote participatory democracy. A critical component of participatory democracy which the management structures aimed to facilitate is decentralization. The next section demonstrates that despite attempts to decentralize functions and services, the promotion of participation and collaboration through representatives led to the 'new centralism'.

Decentralization

This section explores the meanings and impact of decentralization on participatory democracy (in the context of the BEIP). It also explores how the management structure established the policy context for the participatory development processes. In achieving this, it builds on the view presented earlier that retaining decision-making with the taskforce contributed to *new centralism*. It argues that decentralizing functions and services through a management structure in which disadvantaged people participated and collaborated through representatives cut against the aim of enacting the BEIP through the active participation of the selected disadvantaged communities – the disadvantaged people hardly considered the representatives capable of speaking and acting in favour of their interests.

The precise case is that the way the management structures established the context and policy for decentralization actually reduced any possible benefits of participatory democracy. Indeed, it had a defining impact on the way disadvantaged people experienced 'participation' and 'partnerships' under the BEIP. Under the BEIP, decentralization was premised on the view that:

> decision-making is highly centralized at the Ministry of Education headquarters. [In relation to] ... institutional management [decentralization meant] devolving decision-making and resources management to lower level ... structures with the ministry of education [central office] maintaining general oversight and overall superintendence. [It also meant] allowing broad-based participation in the

> provision of education with all stakeholders taking responsibility for planning and implementation. (GOK, 2003, pp. 5, 6, 9, brackets added)

As indicated in this excerpt, decentralization aimed to satisfy the view promoted within participatory democracy that decision-making and functions should not be done at a more centralized level than is absolutely necessary (Ife, 2002).

Based on this view, the government established the management structures used to support the BEIP with the purported aim to support devolution of functions, resources and services from the central office downwards through the district implementation committees and SMCs/BOGs/PTAs to disadvantaged communities. Along with these, the GOK established structures to support broad-based participation and partnerships. Here, decentralization involved sharing management, financial and implementation responsibilities amongst the government, OPEC and communities. The way financial responsibilities and resources were shared and the practices of partnerships and participation the management structures generated are considered in detail in subsequent chapters. For the moment, technocrats' need to retain decision-making authority at the central office cut against effective devolution of functions and obligations to collaborating technical experts, communities and SMCs/BOGs/PTAs. Retaining decision-making authority meant that despite allocation of roles to these groups, the BEIP was still controlled at the central office. This separation of decisions from functions and obligations (or planning and decision-making from implementation processes) contributed to 'new centralism'. Unlike in the old centralism where only technocrats participated, in the new centralism the donor, technical experts, civil societies and the disadvantaged communities played a key role. New centralism also emerged because disadvantaged communities were meant to participate and collaborate in management of the BEIP and educational governance through representatives. Although these sensibilities resonate with the aims of participatory democracy, practice shows that participation and collaboration through 'invited' rather than democratically 'elected' representatives delegitimizes the management structure and contributes to new forms of centralism.

Representative Democracy

As shown in Figure 3.3, technocrats aimed to affect hierarchical and horizontal relationships through participation and collaboration at the national, district and school tiers of educational management and governance. A key strength of these representational formations comes from intentions to enable disadvantaged communities to create social, political and economic networks of change. Ultimately, as noted above, these networks were expected to enable (or it was hoped that they would enable) disadvantaged people to overcome the social exclusionary effects of economic change, bureaucracy, poverty, illiteracy, gender and culture. Thus, the GOK aimed to reconstruct the place of communities to be able to control their own development by providing them with spaces to collaborate and participate in enacting and implementing the BEIP.

The reconstruction of the place of communities in development cannot be overemphasized considering that technocrats assumed the earlier stated structural and rights-based perspectives, and the principles of holism and balanced development where participation and partnerships are core components. A critical examination of how the management structures were meant to provide disadvantaged people through the BEIP with opportunities to reconstruct their own futures shows that technocrats either equated communities with committees or construed management as something communities access through representatives. As shown in subsequent chapters, the paradox is that representation concealed in the management structure the place of disadvantaged people whose lives the BEIP was meant to improve.

By so doing, technocrats relegated the role of active participation by these people in management. To concur with Brown (2004), where people participate through representatives, could be more exclusionary than other forms of management. As discussed in Chapter 5, disadvantaged people and SMCs/BOGs/PTAs were excluded from enacting the most important policies and decisions through which the BEIP evolved. Thus, participation in management through representatives is a form of new centralism.

Participation as a Form of 'New Centralism'

This section builds the case further by arguing that the use of management structures that facilitated participation through representatives negated active participation and benefits of ownership to communities and SMCs/BOGs/PTAs. The GOK chose to enact management of the BEIP through 'invited' or 'appointed' representatives, rather than through more democratic forms of participation, partly because of the speed with which the reform needed to be implemented and partly because of the vastness and diversity of the context of the BEIP.

The BEIP was implemented in schools that represented multiple environmental, socio-cultural,[5] economic and institutional diversities. To attend to the uniqueness of the needs of these schools, participation of a wide range of stakeholders with clear understanding of the cultural and environmental sensitivities was required. An educationist attested to stakeholders' and institutional diversity in the following ways:

> … the degree of being a stakeholder varies from one group to another. In the case of a school, the immediate stakeholders are the parents, teachers, the pupils, sponsor and the government … There is … SMC in primary schools and BOGs for secondary schools. These are representatives of the above people … when these are represented their views are the ones used to implement the project … the wider community is represented … We cannot have 1000 parents coming to make a decision in a school of 1000 kids. The SMC represent them and they

5 BEIP covered the 43 main cultures in Kenya since it was implemented in the then 70 districts (or eight provinces).

> are the ones who selected the SMC and they have faith … that it is going to implement whatever they want … The belief we have is that in democracy, when you choose somebody, what he does, he does it for you. He/she should have consulted … before they say what they say. (Wamsha)

This testimony resonates with Ife's (2002) claim that representative democracy is preferable in large organizations because, even in small organizations, not all members can actively participate. That notwithstanding, representatives must be democratically legitimated by the communities they represent, not by some other authority because democracy entails giving people choices and freedoms. Technocrats assumed that legitimation in the BEIP involved conferment of power onto SMCs/BOGs/PTAs to act on behalf of the communities these represented either through election or consultation processes. As said before, SMCs/BOGs/PTAs are normally taken to represent the interests of parents and communities who elect them. However, the data with technocrats reveal that parents did not choose the representatives who took up management roles in the BEIP.

Instead, technocrats tailored representation in the taskforce, national-district-based committees and SMCs/BOGs/PTAs to suit the BEIP objectives, as detailed later, through processes of 'inclusion by invitation'. Such customization and harmonization by prescribing representatives, though well intentioned, paradoxically reduced legitimacy of the management structure, benefits of ownership and downward accountability (see Chapter 4) to the disadvantaged people. The legitimacy of the management structures to act and speak on behalf of disadvantaged people can be questioned because the representatives were neither democratically (s)elected, nor their roles authenticated through the active participation of disadvantaged communities themselves.

The issue on legitimacy of representatives arose in the data in response to the question *how did you experience representation in the BEIP?* Emmanuel (an educationist) stated:

> I think... representation … has become kind of traditional, that what they say or what is passed at that [national] level is [taken to be] the views of everyone else. In practice that is not so. [We appoint the representatives] … But this representative should not become an authority unto himself/herself. This is a person who is going to represent the views of [communities] ... After representing the views he/she should come back and give feedback to the community because that is what lacks in most cases or consult with the wider community. (Emmanuel)

As detailed in later chapters, such practices attest that many pluralists' and elitists' approaches to participatory development (Ife, 2002) are likely to legitimate the status quo rather than to change it. It also confirms that representative democracy assumes that citizens have power and authority which they willingly transfer to representatives.

Contrary to participatory democracy, representative democracy tends to increase individual power, not collective power, to make decisions. As noted, in practice, many representatives failed to consult the communities that elected them and they tended to speak their own ideas. They also typically failed to give feedback to the communities they purported to serve. In practice, representation thus became a new form of centralism, because the practices of participation and partnerships encouraged upward accountability while negating government responsibility to the people (see Chapter 4), partly because the management structures facilitated such practices.

The above excerpt attests that where communities elect their own representatives through democratic processes their responsibility to ensure that representatives answer to them is increased because they hold the power to delegitimize representatives who do not fulfil their obligations. While the role of participation is to increase such power (Ife, 2002), the GOK's attempt to decentralize decisions, services and functions through representatives actually negated active participation of communities and their power to enforce the responsibility of representatives to them. Paradoxically, representation was meant to facilitate benefits of consensus-building in decisions as the next subsection shows.

Consensus-building

According to the data, technocrats chose to enact the BEIP through representatives because, in their view, it is much easier to attain consensus-building in decision-making in this way than using more direct forms of participation/collaboration. Wamsha said:

> My experience is that consensus becomes difficult when participants are many. If you are to call a school of 700 children you may have 500–600 parents. If you are to call these 600 parents to come and decide whether to build a toilet or to build a class or administration block, it will be more difficult to reach consensus. And at the end of the day, you may not agree on anything because the views are as varied as the numbers. But when you reduce the people by representation at least you are now drawing towards consensus. (Wamsha)

Consensus-building is an important element of participatory democracy (and/or deliberative democracy in cooperative decision-making) (Ife, 2002) that technocrats aimed to achieve through the needs assessment processes. The data show that technocrats were not unaware of the possible deleterious effects of representation upon participation. Despite awareness that participation is a democratic right, and that more direct forms of participation increased benefits of consensus-building, technocrats chose to use representatives to save time. To this view a technocrat avowed:

> I think representation is about democracy. Even in the political arena, when you elect a member of parliament, he's supposed … to represent your interest … but what he does in parliament is a different story … the same can be argued in these other areas … but the idea of representation is noble in the sense that we cannot all take part in certain things [management] … a lot is wrong with representation. The

> bigger the consultation the better normally, but again, we are constrained by time, and all those things. How we can get out of that is an issue. When I am elected as a representative I stand for corporate benefits but what we see many times are individual benefits. The whole issue of representation … digress because I am there on my own now, not on behalf on the people. (Duncan)

This finding concurs with critics (Cook and Kothari, 2001) who argue that in the name of consensus-building, decisions in mainstream participatory development are rushed to satisfy donor demands and personal interests. The next section shows that the reduction of the democratic right of participation (participatory democracy) into representational roles which increased individual, not collective power, diminished benefits of ownership.

Ownership

The promotion of participation and partnerships through representatives cut against ownership and decisions, participation and freedoms of SMCs/BOGs/PTAs and disadvantaged communities. SMCs'/BOGs'/PTAs' testimonies show that representation limited rights, freedoms and benefits of ownership in the following ways:

> … as a stakeholder … my [head teacher's] view is that, even if we were represented, it is like our participation was minimal because after the representation, there were particular guidelines that were laid out as per how the project is to be implemented … Even if we would like to voice our views, these guidelines, sort of limit us. We get limited because we are told this is what you should do in the implementation of this particular project. (Benjamin)

Ownership entails control of decisions, resources and processes by subjects of development (or disadvantaged communities) (Ife, 2002). This was not the case in the BEIP. In the context of the BEIP, representation became a further form of centralism and, as detailed under consciousness raising and capacity-building (Chapter 5), a mechanism of domination and exclusion. Although technocrats devolved financial functions to the grassroots to enable SMCs/BOGs/PTAs to construct classrooms, laboratories, and water tanks, and provide other services targeted in the BEIP, these management committees barely possessed sufficient authority to vary the decisions technocrats made even where such decisions did not satisfy the unique needs of the specific schools. Again, in practice the management structures introduced by technocrats enacted the BEIP without active participation of disadvantaged communities. The policies and decisions technocrats made also impacted upon practices of participation and partnerships in ways that ignored important contextual and environmental differences. For example, the ceiling costs that defined the total amount of aid each school community was entitled to limited freedoms and alternative choices of SMCs/BOGs/PTAs and the advisory roles of engineers who offered technical expertise on constructions. One head teacher (secretary to a BOG) expressed his concerns as follows:

> The public works officers were supposed to guide us in working out the bill of quantities. Again, they are restricted to the ceiling costs and the amounts allocated for that project. So even if they would work out the bill of quantities and recommend this is it, they could not move out of that ceiling … ideally all projects wherever they are located they were put on the same cost index. So there is little that we can do about it. We were guided and fixed within that bracket. (Hamish)

These guidelines were perceived to impact 'universalising' effects that effectively negated contextual differences. As such, representation became a form of new centralism because it excluded SMC/BOGs/PTAs and the communities from enacting the BEIP in the way intended. Indeed, it devalued local potential knowledge and left inequalities of gender, class, poverty and culture unaltered. As Ife (2002) argues, such practices are what the management structure ought to have challenged through participation and partnerships.

Despite the need to affect bureaucracy and its undemocratic practice, participation and partnerships through representation turned out to be a mechanism for reinforcing the status quo through means of domination or coercion. The way technocrats dictated resources and methods of participation is recorded in Wamsha's testimony:

> There is a lot of participation of the community … After we come up with what we want to do and that one is determined by the donor and the government. Then the means and ways of implementing what is to be done is actually done by the stakeholders … For example … when *OPEC and the government decided* to construct five schools in every district. From there the government had no more say. It only *dictated the budget*: that the budget is 2.1 million for every school ... from there we trained the head teachers, the sponsors and accountants so that they can know how to handle the project. From there the ministry has been relying on the same stakeholders to implement the project. I think we are doing well. All were involved no body can say this project belongs to the government or to OPEC. Apart from the money only, the other bits come from the stakeholders [parents/communities]. *There are a few things like the [construction] plan, the sketch, the specifications which have to be controlled by the government* … but you see *stakeholders cannot provide that.* But when it comes to the real work, it is the community.

While there were good intentions to involve BOGs/SMCs/PTAs and the communities they represented in decision-making, the use of representatives concealed power differences (Chapter 5) and legitimated the status quo. The legitimating role of representation was recorded. Wamsha observed that SMCs tended to lose their critical advantage once they began to enact their management roles:

> It is funny because the SMC has been selected by the same parents but to some extent sometimes they feel as if they are part of the school administration, they are not representing them [parents]. I don't know how that one develops but my experience is that at some point you hear some parents accusing the same SMC for rallying together with the school administration, which should not be the case … [the wider community now hold representatives] with suspicion as if they are not representing their views.

This idea of representatives 'rallying together with' (or uncritically supporting) school administration practices, which they hitherto ought to challenge underscores two competing principles of representation: cooperation and competition. According to Ife (2002), governments and change agents should facilitate democracy by enacting policies and methods that support cooperation, not the competitive ethic that is to date dominating development. As confirmed here, many technocrats viewed the institutions of SMCs and school administration as two competing, not cooperating, entities. This reading of SMCs/BOGs/PTAs and the communities they represented meant that representation was expected to diffuse competition by reducing the number of participants who undertook management roles. Ultimately, this was seen to generate cooperation and consensus-building in decision-making by reducing multiple views that would arise were they to use direct forms of participation. It is no wonder then that SMCs/BOGs/PTAs lost their critical advantage and began to support their own interests or practices which did not attend to the needs of the communities they represented. As a result, parents became suspicious of SMCs'/BOGs'/PTAs' intentions, thereby both became passive participants in decision-making as confirmed in what Wamsha termed a three-man-show:

> [during] the implementation … the SMC in the schools I have visited... has become a three-man-show. It is drifting from the whole committee … to … few people: the head teacher [secretary], the chairman of the school [committee, who is also] the chairman of the tendering committee … [and the treasurer] … The other members of the committee … claim to be aware and to be involved but in the real sense the real involvement in very minimal … [the wider community involvement] is [also] minimal [in making decisions] … in terms of implementation, supervision and any other issue to do with what is being implemented.

It is no surprise that parents perceived SMCs/BOGs/PTAs to be lacking in their collective synergy to speak on their behalf. The reduced number of SMC members who participated in decision-making shows that participation through representatives evolved into new centralism. New centralism also arose from the principles and approaches used to support development cooperation and partnerships. As detailed in Chapter 4, aid assistance, technical expertise and representation led to the emergence of elite-to-elite networks in which the disadvantaged people were excluded. This way the processes of 'inclusion by invitation' through which technocrats recruited representatives contributed to new centralism as shown in the following subsection.

Inclusion by Invitation

As detailed later, technical expertise, representation and aid assistance underpinned the decision of technocrats about whom to include in the BEIP partnerships. These components represent 'inclusion by invitation', not participatory democracy. Inclusion by invitation meant that only those who offered aid assistance, technical expertise and were assigned representational roles in the management structure were directly involved. The practices of participation, cooperation-partnerships born of these features, reinforced bureaucracy and its undemocratic practices. Carla (technocrat) said:

> Sometimes it [bureaucracy] can hinder in the sense that if you are to discuss something, it is not like they [National Taskforce, Steering Committee, donor] have all the time to sit with you [project implementation unit] and discuss. Maybe they are out on a meeting … [or] there are certain things they want to be explained or harmonized. By the time that information reaches you so that you can go back, maybe another week is gone.

As shown in later chapters, bureaucracy is a significant inhibitor to cooperation, participation and empowerment. When asked to reflect on an ideal structure that would help reduce such delays and red-tapes, technocrats and SMCs/BOGs/PTAs expressed the desire for shorter bureaucracy with fewer management structures at the national office. Their testimonies concurred with policy documents (GOK, 2005a, 2005b) in which it is suggested that decisions were decentralized to implementation units that dealt directly with the disadvantaged communities. Paradoxically, though acknowledging the deleterious effects of bureaucracy, some technocrats did not see the need to decentralize decision-making power further down beyond the project coordinator to the districts or the school-based structures. A key problem arose from technocrats' views of bureaucracy as only being important in development management inasmuch as it hindered cooperation and delayed project activities, not in limiting rights (self-determination) and freedoms (association) which were core concerns for SMCs/BOGs/PTAs and some district implementation units. Their suggestions were still forms of centralism as a parent/BOG member illuminated when she said:

> When you talk about the GOK, I think participation is far from anything … people in the ministries … don't seem to understand participation. They are operating in a structure that is so *top-down*, that the new concept cannot be conceptualized. All of them are getting directives from above. To go and sit with the community and tell them to participate to come up with anything, it is completely abstract ideas based on their context … While many people in the mainstream government and NGO world will basically say participation is good, the practice of participation is quite far from … reality. NGO world have a better understanding of it but for their own selfish reasons, and for the fact that very

> few want to share *power*, where it means let us all with the *beneficiaries have same knowledge* in everything, share the knowledge/the information concerning the project so that they can make informed decisions, they don't want that … but then the government … is even worse because even the awareness and the context does not embrace participation. (Sundukia)

These findings confirm criticisms that government-donor-led [participatory] development can easily slip into neoliberalism, since it cannot shed off its modernist past (Manzo, 1991). By establishing a hierarchical management structure, technocrats reproduced in the BEIP the culture of hoarding information that was dominant in the hierarchical structures of the education system. Again, technocrats' proclivities towards representative democracy reinforced top-down mindsets which limited the benefits of participatory democracy. As detailed in later chapters, representation, technical expertise and aid assistance, as established in the management structures, obscured benefits of empowerment and social change to disadvantaged people.

Conclusion

This chapter has argued that the management structure through which the BEIP was enacted had a defining impact on the way the people involved experienced and perceived participatory development and participatory democracy. Key strengths of the BEIP structures emerged through the use of holistic, balanced, structural and rights-based perspectives. The integration of participation and partnerships into objectives with a view to increasing democratic practices and reducing the social exclusionary effects of bureaucracy, poverty, culture and gender and the focus on empowering disadvantaged people to control their own futures represented a political will to promote sustainable livelihoods. However, the way technocrats formed the management structures established a fertile ground for the emergence of bureaucratic technical expertise and aid powers, which were responsible for exclusion of the disadvantaged people in decision-making processes. Indeed, the convergence of the perceived powerlessness of the poor with these three types of power was meant to establish an environment conducive to empowerment of the disadvantaged people. Nonetheless, the use of a management structure in which disadvantaged people participated and collaborated through representatives who were not democratically elected reduced the benefits of participatory democracy and led to the emergence of new forms of centralism. The practices of participation and partnerships, though imbued with the potential, barely unleashed possibilities of change (or 'power to'), cooperation/collaboration on an equal basis (or 'power with') and feelings of self-esteem and self-worth (or power within) amongst those involved. Questions can thus be raised about the extent to which partnerships and participation enhanced empowerment, sustainability and social change to the disadvantaged people. The next chapter explores the extent to which the BEIP contributed to the formation of partnerships on 'equal bases' and their impact on accountability.

Chapter 4
Development Cooperation, Partnerships and Accountability

Introduction

As established in the previous chapter, the GOK created representative committees at the national, district and school levels. These committees were created to increase the role of disadvantaged people in educational governance and management. The view was that the committees would increase collaboration and participation and provide spaces for disadvantaged people to form durable socio-economic and political networks. In turn, it was hoped that these would promote development that was truly sustainable.

This chapter examines further the extent to which the policies, approaches and principles used to facilitate collaboration actually enhanced sustainable development. It draws upon policy documents and empirical data to offer a critique of the efficacy of the BEIP to achieve its goals. It also argues that the partnerships established under the BEIP were premised on the need to maximize aid benefits through representation and technical expertise. In practice, however, donors and elites were empowered at the expense of the groups and individuals the BEIP was designed to help. Indeed, the data suggest that power shifted towards elites and the donor, not the disadvantaged people.

The structure of the chapter is as follows. The next section looks at the policies, aims, approaches and principles used to create partnerships in the BEIP. The central argument is that partnerships drew upon market principles. Besides negating benefits of equal partnerships, this contributed to emergence of aid and technical expertise as power. The following section interrogates processes of collaboration in planning, implementation, and monitoring and evaluation activities of the BEIP to establish the nature and impact of the emergent partnerships and social networks on cooperation and accountability. Here, it is argued that the partnerships generated through representation, aid assistance and technical expertise reproduced elite-to-elite networks, enhanced competition – not cooperation – and facilitated upward – not downward – accountability. In conclusion, a summary of the weaknesses and strengths of the BEIP partnerships is given.

Principles of Development Cooperation and Partnerships

As described above, the BEIP was implemented through a government partnerships involving the Kenyan Ministry of Education, OPEC selected school communities. For its part, the GOK hoped that the selected communities would access aid assistance from the OPEC through *representation* by stakeholders (GOK, 2003b, 2005a). The core stakeholders included technocrats (educationists), technical experts (from the Ministries of Water, Public Works and Housing and Health) SMCs/BOGs/PTAs, and the disadvantaged people.

The current impetus within this genre of partnerships in the education sector comes from the policy on sector-wide approaches to planning (GOK, 2005a, 2005b). A sector-wide approach to planning is:

> … a process of engaging all stakeholders in order to attain national ownership, alignment of objectives, harmonization of procedures and approaches and a coherent financing arrangement. [It] involves broad stakeholders' consultations in designing a coherent and rationalized sector programme at micro, meso and macro levels [school, district, national] and the establishment of strong coordination mechanisms among donors and between donors and the government. (p. iii)

Sector-wide approaches to planning require partners (or stakeholders) to consolidate resources into one 'basket'. This enables governments to perform more effectively their roles of coordination and facilitation. In this way, sector-wide approaches to planning underpinned the aims, meanings, processes and the anticipated outcomes for creating and promoting partnerships, cooperation and networking with other projects (or organizations) that shared objectives with the aid interventions, such as the BEIP.

Notions of partnerships vary according to the communities that were involved. However, in the main, these echoed participatory development practices of CBOs, women's groups and other self-help groups (Chambers, 2005). Interview data gathered as part of the present study attested that while the idea of sector-wide approaches to planning found in the BEIP echoed harambee and indigenous principles of participation and partnerships, integrating such ideals in policies expanded the scope of partnerships from family/village/community to include national and international actors. The data also showed that the Kenya Education Sector Support Programme was formed through sector-wide strategies and that all its 23 programmes ought to be implemented this way.

The infrastructural programme of the Kenya Education Sector Support Programme is a conglomeration of projects, which the GOK is implementing in partnership[1] with multilateral and bilateral donors, private sector, media and

1 The infrastructure support for North Eastern Province primary schools funded by GOK/USAID; Arid Lands Resource Management Project funded by GOK/World Bank;

civil society (community, NGOs and religious organizations). As one of these projects, the BEIP was central to the Kenya Education Sector Support Programme partnerships. Although at the time the sector-wide approaches to planning policy were enacted, implementation of the BEIP was already underway, and documentary and empirical evidence support the view that technocrats encouraged *collaboration* and *networking* between the BEIP and these broader partnerships. The BEIP cooperative approaches also drew upon the economic development policy of the Poverty Reduction Strategy Paper and Economic Recovery Strategy for Wealth and Employment Creation.

These sensibilities aimed to balance global and local perspectives of cooperation and partnerships. They conformed to Klees' (2001) view that a genuine sector-wide approach to planning must:

> ... be sector wide; based on a clear and coherent policy framework; local stakeholders are supposed to be fully in charge; all main donors must agree to it; implementation must be developed jointly; and it should depend on local capacity, not technical assistance. (p. 112)

At this juncture, it is reasonable to ask to what extent the practice of development, cooperation and partnerships in planning the BEIP satisfied these criteria. Contrary to cultural practices of participation and partnerships that were people-centered, the data show that the whole idea of 'pooling resources into one basket' in sector-wide approaches to planning (as experienced under the BEIP) redefined these features in ways the led technocrats to focus more on promoting development through aid assistance rather than on developing local capacities. The focus upon development through direct financial aid limited the extent to which the BEIP promoted sustainable development. In practice, pooling resources in the BEIP was taken to mean harmonized functions and services through creation of structures of collaboration and participation. The view is that partnerships and participation would create new socio-economic and political networks based upon cooperation. As noted above, this satisfied the broad criteria that participatory development should create 'political spaces' for the hitherto excluded to participate and collaborate (Cornwall, 2004). It also conformed to Ife's (2002) view that as an essential component of empowerment and social change, cooperation enables disadvantaged people to influence structures that govern their lives.

However, while in theory the BEIP was consistent with the need to increase collaboration and partnership (particularly with disadvantaged groups), the data suggest that the practice differed. When establishing the BEIP, the Steering Committee and Project Implementation Unit, held discussions with semi-autonomous government agencies of the Ministry of Education and UNESCO (GOK, 2002b). These initial discussions developed the project proposal to meet

---

Community Development Trust Fund; KFW support to primary schools in all areas; Local Authority Trust Fund (LATF); and Constituency Development Fund (CDF).

the structural and rights deficits of disadvantaged school communities identified before. By developing the proposal through which they fundraised to support the BEIP, technocrats also determined the areas of need and the ways in which communities should collaborate and participate. Although this practice appeared to satisfy sector-wide approaches to planning criteria that 'local stakeholders must be *fully* in control', it contradicted the principle that change must come from below. One way that the practice had this effect was by excluding disadvantaged people and SMCs/BOGs/PTAs from enacting such plans and decisions. Such exclusion is also evident in the negotiations that followed between the Ministries of Education and Finance (represented in the national taskforce) and OPEC. The project loan agreement and the proposal upon which the BEIP was based (GOK, 2002b) show that these negotiations established the nature of the BEIP partnerships by defining the collaborative processes/roles of the donor, technocrats and SMCs/BOGs/PTAs. As Carla put it:

> We [technocrats] have given them [OPEC] our proposal. They have accepted and are willing to fund the construction of those facilities [classrooms, water tanks, toilets] to the tune of 2 million [Kenya Shillings] per school ... The donor[s' role is] ... giving the financial [assistance] and … the conditions on how that money should be used. So we discuss together and see if those conditions are applicable. So their role really is to hear from us and we also listen to them. They say this is the amount of money. We say the best way we can … administer this money is … They may also have their own suggestions. For example, when we started OPEC [project], they were wondering whether we should have a district tendering or school community tendering process. We discussed together and came to an agreement that tendering should be done at the school level. So, their role is to give financial aid and also advice … We share ideas about how the money should be administered.

The chapter will return to these conditions later. Given failures of previous aid projects to achieve *sustainability*, it was established that the BEIP would be implemented through the active participation and collaboration of the donor, school communities, and technocrats. Here, development cooperation and partnership meant sharing management and financial responsibilities between the key stakeholders (GOK, 2003b). However, when asked to reflect on the nature of cooperation in the negotiations which enacted decisions to implement BEIP through partnerships, Benjamin, a head teacher said:

> I think in the first stage, they [technocrats] … initiated the idea of bringing stakeholders together to get engaged ... they prepared project documents … The plan the project was to take was according to the *test* not only of the Ministry but also of the donor [including] signing the agreement with the donor.

In practice, then, the privileging of personal and organizational interests in aid and markets raises questions about whether or not technocrats, donors and civil societies actually 'represented interests' of disadvantaged people through the BEIP (Ife, 2002). The use of technocrats, political elites and technical experts to promote participation confirm criticisms that government-donor-led development equates the nation-state with the political subject (Manzo, 1991). Although disadvantaged communities were the real political subjects, neither them, nor the SMCs/BOGs/PTAs (who were meant to represent their interests) initiated and negotiated the BEIP with funders. Technocrats assumed that their interests and those of political elites, donors and technical experts were synonymous with, and represented, those of disadvantaged people. Yet, as a component of deliberative democracy, the aim of cooperation and collaboration is to decentralize decisions and services from the central office to the grassroots level through active participation. In practice, however, under the BEIP, the data show that excluded groups floundered to identify with, and own, the decisions emerging from such negotiations. Thus, technocrats' view of 'partnerships as sharing responsibilities' to redistribute costs and maximize benefits of aid, privileged aid assistance over local potentials. This negated creation of partnerships on an equal basis and led to emergence of aid as power.

Aid as Power

This section explores the themes of partnerships as sharing responsibilities and 'partnerships (and participation) as conditionality'. The central argument is that the principles that underpinned partnerships which were established under the BEIP led to a shift in power away from disadvantaged communities towards donors. Technocrats played a key role in facilitating this shift.

As noted above, the GOK in collaboration with the donor established the amount of aid assistance each school would receive. They also agreed to the formation of school-based tendering committees and established procedures of contracting and procuring goods and services/labour through competitive bidding that these committees were required to use. Antoinette described the processes of negotiation, nature of intervention, contexts, conditions, policies and processes of collaboration in the following ways:

> One thing I admit with OPEC is that it [loan] was negotiable. They allowed negotiation and renegotiation … I realise there is no decision which OPEC came forth with, which was not giving an opportunity for understanding. We sat quite a number of times to review the Aid Memoir. Through it, I realised that some of the things which were in the agreement were revisited to make them workable. For instance, there was this time they [OPEC] wanted to send the money and they wanted that money to come in a lump sum. We said no. We want a revolving fund. That is, money sent to the Central Bank we use. When it is over, then they replenish what we have used. Those were renegotiations … I

> also remember they had wished that when it comes to reporting back or giving feedback to the donor [accountability], they had said we make a report for the whole country once. We said no. It has to be in phases. Another better example is when it came to contractors. They had wanted us to get one contractor for all the schools in the country. We sat down and said no. We cannot afford the risk with one contractor. Let each school manage their own contractor. All those things were renegotiated. Despite all these, I think OPEC was quite good because they allowed room for renegotiation.

Negotiation is critical in establishing and maintaining partnerships. Negotiations at the planning level offered 'space' for the donor and technocrats to exchange ideas and agree on the policies, procedures and organizational structures that would be used to implement and manage the BEIP. Consistent with Ife's (2002) view that issues of networking are better handled at the lower tiers of government administration, technocrats established procurement, tendering and contracting committees at the school level. Technocrats also established policies and guidelines to support decentralization of services and responsibilities from the national office to these grassroots (or school-based) structures. The aim was to increase the role of SMCs/BOGs/PTAs in management and control of the aid finances and other resources offered to their respective school communities.

Some technocrats also believed that decentralization would increase the participation and collaboration of the disadvantaged communities in the BEIP and thereby promote the creation of partnership and social networks across social classes and on an equal basis. Ultimately, these features would enhance government accountability and responsibility to the broader disadvantaged communities. We shall return to the extent to which the emergent partnerships and social networks enhanced accountability later.

The data also show that while the policies and guidelines supported collaboration and equality, the actual practice of negotiation negated formation of partnerships on an equal basis. An important reason for this was that the groups and individuals the BEIP was designed to help were excluded from the initial negotiations. As said earlier, this happened because, despite understanding that grassroots structures would implement the BEIP, collaboration primarily occurred between technocrats. These technocrats were believed to possess the knowledge and skills to negotiate the loan, mediate the inflexibilities of donor funding and make the decisions and terms agreed upon sensitive to the realities of the disadvantaged people. Antoinette, reflecting on the BEIP negotiations, observed:

> There are some donors[2] I know but let me not mention their names. I think why they do not want (re)negotiations is because they have some vested interests … So this is why they may want to be so rigid on what they said from the word go.

---

2 Refers to the development partners in the Kenya Education Sector Support Programme and the infrastructure programme cited before.

> But if one is out to help, they should help and help indeed. Their rigidity is out of self-centred gain … [Also] this is a loan and not a grant. There are those who are giving grants and they have to dictate and you [borrower] because you are a consumer and helpless, you just go by their terms. You can't change, because it is a grant after all … I believe in every project if it has to bear fruits there should be understanding.

Although the OPEC allowed for (re)negotiations, the exclusion of disadvantaged people and the conditions which the donor and technocrats agreed upon compromised formation of partnerships on an equal basis and led to emergence of aid as a form of power. In this context, aid was power because it gave donors leverage to engrave conditions that constrained the ability of disadvantaged groups to make their own decisions. These conditions also accentuated vulnerability. Both theory (Ife, 2002) and practice (as shown in the educationist's comparison between the relational dynamics that emerge based on whether donors are providing a 'loan' or a 'grant') tell us that disadvantage limits the choices of the government and disadvantaged people. Again, donors imposed conditions even where opportunities for negotiations were created because they found it hard to change their ways of thinking and doing things, as Mapatano, a councillor, confirmed:

> We have met NGOs and many international organisations which come and partner with the government for specific projects. When these people come, they have a specific line of operation. You cannot deviate … it is like if you ask World Bank to change their way of doing things. You cannot ... They will make sure that you are put in their way of doing things and think the way they think ... I think in OPEC, every beginning can be the beginning of our design. We [can]not talk of planning or negotiations as our initial design level … We talk at the level when money is available and we need to be called to participate as our beginning.

These data show that the inability of donors to change their ways of thinking and acting, and, critically, the inability of the GOK to effectively influence the decisions and intentions of donors, limited the effectiveness of the BEIP. These inhibitors of effective partnerships lend support to Brown's (2004) view that most communities construe donor and government institutions as distant from their realities and possibly acting at odds with their needs. For example, the inability of donors to change mindsets led to conflicts of interest and, ultimately, to the withdrawal of the BEIP from primary schools in Coast Province. The conflicts of interest occurred because the World Bank wished to implement a different project in the Coast Province where the BEIP was already being implemented. Since the GOK needed to ensure equitable distribution of the projects supported by donors through the Kenya Education Sector Support Programme, the BEIP was removed from such schools to pave the way for the World Bank (which could not change its focus) to implement its preferred project. In addition, such practice encouraged competition among donors to advance their preferred projects, without due regard to the participation and collaboration needs of the disadvantaged people.

The GOK also established structures that SMCs/BOGs/PTAs described as conditions (or blueprints) as the words of a teacher reflected:

> There were guidelines, or blue-prints as well as the relationships … we [SMCs/BOGs/PTAs] could not go beyond those blueprints … (Hamish)

The participants viewed the structures that the GOK established to enact partnerships, participation and accountability as *conditions* because they prescribed these features in ways that arguably limited full participation and freedom of choice. More significantly, the interviewees stated that donors and technocrats imposed conditions to protect their own interests in markets and in aid programmes. Although claimed to enhance good governance, the data show that such practices also created mistrust where aid funds were disbursed in instalments. In this respect, Mapatano reported:

> … it is like the donor or whoever is giving aid, number one does not trust the person he is giving [the aid to] … He is ready to give but … that donor money … is also given some conditions … In this OPEC, what we are seeing is that, 'we shall give you five hundred thousand first. Clear that … before that … we cannot give you another [amount] … account for that … [then]. We give you another one' [as set out in the revolving fund] … forgetting that there is timing here where the contractor is waiting [to be paid]. So one week being wasted, that is going to be paid for and there is no work going on. It is like the donor and whoever is giving doubts the person he is giving. Whoever is giving does not tell us I am giving you this much, this is what I had … planned for you. You are just given, it is like somebody's own property and [he] lives the life for you.

The GOK and OPEC decisions to implement the BEIP in three phases and to disburse funds in instalments, as shown in the revolving credit above, although it is what technocrats wanted, had the undesired effect of reducing the confidence of disadvantaged people about government and donor commitment to enact structures that affect sustainable changes. SMCs/BOGs/PTAs testified that as a result of these decisions (conditions), their freedom of choice and ability to plan ahead was limited. Rather than enhance cost effectiveness, the policies that supported disbursement of funds in instalments also encouraged delays in the procurement and delivery of services and goods. As a result, the BEIP could not be completed within the desired timeframe.

As the political councillor testified, these conditions could be seen to enable donors to encroach on disadvantaged people's lives while enabling them to mistrust the democratically elected institutions of the government, instead of enhancing cooperation amongst all. Indeed this revelation adds weight to Brown's (2004) contention that partnerships based on aid concessions diminish government legitimacy and distort authentic political relationships. This point will be developed under accountability. For the moment, it is useful to note that the imposition of conditions based on aid concessions potentially reduced the participants' confidence of the GOK, since it represented donor and technocrats' knowledge, as the unquestionable norm (even where it further marginalized disadvantaged people).

Technical Expertise as Power

As alluded to earlier, technical expertise is premised, in part, on the view that disadvantaged people do not understand what is in their best interests. Even where disadvantaged people understand their entitlements, they usually lack resources, technical knowledge and skills required to meet these. These resources, knowledge and skills gaps, it is claimed, must be filled by experts. The enactment of partnerships through aid assistance and technical expertise led to emergence of technical expertise as power (of resources and knowledge). Technical assistance as power was recorded where Bushie, an educationist, said:

> … there is this assumption that developing countries need technical assistance. So they [donors] come and say 'we want to help you, what can we help you with.' Once you say, 'yes' to the project, half of the project money goes with their salaries. It is a way of creating jobs for their people. At the end of the day, only a ¼ of say the 15 million the donor contributed is used on project implementation. You don't feel the impact … but they will write that they assisted with 15 million, when only 2 million reached [the community]. Other donors want to maintain off-shore accounts. It is them to decide which and who to contract, select consultants, decide how much to pay … So, they end up saying we gave you 15 million yet only 8 million reached you. The rest was used elsewhere because of donor interest to create jobs, decide and control the process. So now it becomes a problem with these donors.

This confirms power and market relationships in aid partnerships (Kapoor, 2005) where disadvantage (or need) constitutes the 'power(lessness)' upon which technical assistance thrives. Technical assistance can be a form of power because partnerships are seldom 'mutual'. In this context, technical assistance negates equal partnerships because it is imposed as conditions. In other contexts, technical expertise created mistrust. For example, to entrench their interests, head teachers were reported to have used their superior technical expertise to compromise the less educated members of their communities:

> I want to state that sometimes the suspicion [of communities] is true. Those people sometimes are compromised ... If you go to the rural areas, you find that some parents who are in the SMCs are actually illiterate and are actually the

> grandfathers of the children in that given school. The real parents are not there. Grandfathers are not the ones who give the money. They may have very little information on construction. So they may actually be the only people [available] in the whole community. So it is very easy for the head teacher to compromise, if he wants to comprise them … most of them are actually compromised. To reduce such suspicion, let us have SMCs that consist of literate people. I think Kenya now we are in a stage where we can say any person who is not literate should not be in any given committee. (Wamsha)

For their part, some other technocrats believed that SMCs/BOGs/PTAs lacked the literacy levels necessary for leveraging criticism to head teachers. Technocrats also considered that some SMCs/BOGs/PTAs comprised aged people who were incapable of providing the technical expertise needed for constructions and for effective management of the BEIP. Again, since the BEIP used pre-existing structures at the grassroots, members who undertook management roles were neither the ones experiencing the needs (at least directly), nor those raising the resources used to build the physical facilities. SMCs/BOGs/PTAs were considered devoid of the capacity to effectively manage the BEIP and to enforce accountability of school administration to the GOK, nor responsibility to parents. For reasons such as these, technocrats considered SMCs/BOGs/PTAs incapable of managing the BEIP effectively. Since SMCs/BOGs/PTAs could neither enforce head teachers' accountability to the Project Implementation Unit and answerability to parents, the embryonic partnerships in the BEIP were seen either to accentuate inequalities, cement the status quo, or provide a fertile ground upon which pretentiousness in the discourse of ceding power to the disadvantaged people continued to thrive.

Indeed, as expressed in the BEIP, such partnerships contradicted the view that technical assistance should facilitate the empowerment of all (Arnstein, 1971; Ife, 2002). The problem was that technical assistance allowed donors and technocrats to offer solutions according to their own perspectives and interests. Such solutions were not typically feasible and, as we shall see in the next chapter, devalued local resources and knowledge. For example, in one instance, technocrats purchased ten vehicles to support activities of the BEIP. As Bushie said, such vehicles were incompatible with some local terrains and " … may not end up improving the general wellbeing of these so called people to be assisted".

The paradox is that the structures created to address these deficits facilitated elite-to-elite networks, since they drew upon technical expertise, and, as the example above shows, helped establish partnerships that were not sustainable. Arguably, the associated exclusionary practices exacerbated the situation. For example, the GOK prescribed that members of the SMCs/BOGs/PTAs must at least have literacy levels of primary or secondary education to assume management roles. As Antoinette said:

> We did not ask for uniform level [of literacy] because it depends on the regions, but mainly in the SMCs we expect someone who can read and write, because if we put the levels of education we will block those who are from marginalized areas and yet they are not learned as such. But our preference in secondary education is that we get someone who has higher education.

Thus, technocrats fixed literacy levels for members of SMCs/BOGs/PTAs so that it would be easier to communicate with them regarding the coordination of the BEIP processes. As said under decentralization, technocrats viewed SMCs/BOGs/PTAs as 'gate-keepers' (Eade, 2003) through whom they could access disadvantaged communities. Despite some technocrats' concern about the need to avoid exclusionary structures, the policy on minimum literacy levels for SMCs/BOGs/PTAs actually explicitly excluded the less educated folks from management roles.

Moreover, the practice of 'sharing responsibilities' and 'harmonization of activities' through representatives, aid and technical assistants contributed to emergence of elite-to-elite networks and partnerships through *competitive* rather than *cooperative* relationships. Cooperative relationships were meant to emerge as a result of the BEIP. For example, when planning how to implement the BEIP, technocrats collaborated with SMCs/BOGs/PTAs to assess the needs of schools. However, data show that in practice, such 'collaboration' privileged technical expertise and excluded ideas of SMCs/BOGs/PTAs in the final plans that these grassroots structures were meant to implement.

> … I remember, I was in the first meeting. We were able to come up with some suggestions. Do we need a new laboratory or do we renovate the old one? Based on our agreement … the plan was drawn. To some extent we were involved but the *technical expertise* was to be done by the Ministry of [Public] Works. (Hamish)

> Incorporation of ideas and suggestions of science teachers in the planning of the project was not done. Plans were just brought in and we were told build according to this plan. So we had no alternative but to follow what they had decided. (Benjamin)

The failure in the BEIP to incorporate local interests lends support to the views of critics of mainstream participatory development who argue that partnerships within aid delivery systems continue to perpetuate domination (Ife, 2002). As noted, a key way this occurred was through the role played by the BEIP in supporting the formation of elite-to-elite networks. A further way the partnerships undermined participatory democracy was through promoting competition. This issue is discussed more fully in the next section.

Competition

Interview data recorded conflicts of interest, competition and duplication of resources/services in the partnerships that emerged through planning, implementation and joint monitoring and evaluation activities. One educationist said:

> There are times when we have duplication of activities. You find that everybody is targeting educational programmes … Early childhood development, girl-child education, government and aid-sponsored programmes … You [also] find that one NGO is doing the same [thing] as another … There is … overlapping in the activities. But to ensure that there is no conflict we advise them to work in different areas … There are a number of times when they do a similar activity ... [that brings] conflict of interest … in the sense that … One has decided to do one thing which is similar, and you are doing that one thing in the same community. That duplication is there. Conflict of interest in the sense that ... many times there is a bit of friction … because everyone wants to serve and want to be seen serving the community. Everybody wants to remain in the community to ensure that when their evaluators or monitors come they are given a good report … so those things are there. (Dorobo)

A contributing factor to conflicts of interest and competition arose from the use of representatives and formation of partnerships on the basis of aid assistance and technical expertise. These features heightened individual rather than collective power and led donors, technocrats and NGOs to take advantage of the conditions of disadvantaged people to justify their own material interests in aid:

> These stakeholders [donors, technical experts) come and assess our needs … They want to ensure that they narrow the gap … They help us at least … rise to certain levels with the rest of the [developed parts of the] country … The idea is to ensure that we have risen from that low level where we are, that unprivileged place … because of the geographical position of those communities. They are very far [from] these parts that are considered developed and having all the opportunities. So we very much appreciate these stakeholders and the idea is to ensure that we come up … Other motives [are] … most of these stakeholders [government and NGO] are people who get donations from outside. Of course their existence … depends on the activities that they generate within certain areas. So as much as they want to help us and that seems to be the objective, they also want to ensure that they exist and of course get money using now the communities' … poverty level [condition]. So that proves to be an excuse of all of them … for earning money from donors. (Dorobo)

Contrary to the BEIP's aim of enhancing cooperation, technocrats' need to justify personal and corporate interests in aid and donor interests in markets led to the formation of partnerships based upon competitive, rather than cooperative relationships. For example, in Western Province some schools participated in a pilot study in which technocrats tested efficacy of the tools and methods associated with the BEIP. Interview data show that the BEIP was not actually implemented at all in some of the schools that took part in the pilot because the selection processes involved politicians whose interests mediated against cooperation. While there were no conflicts of interest between the donor and politicians, their cooperation did not necessarily work in favour of the collective good. For example, the multiparty political context undermined cooperation because some of the members from opposition parties could not readily identify with the policies and decisions of the ruling party. Such milieus encouraged competition between the interests of politicians and those of technocrats some of whom succumbed to the whims of politicians rather than go by the selection criteria discussed earlier. This conflict of interest arose because political elites:

> felt they needed to influence our [technocrats'] decisions on where to implement the project. So after we took our stand and we insisted on the areas and the criteria that we gave and stuck to it, they realized that since this is a benefit to their constituency, they had no choice but to support it. So we have got a lot of support even though the project is mainstreamed within the government, whether the politician is on opposition or even the government in power … (Antoinette)

The attempt by technocrats to privilege objectives of the BEIP (which supposedly were taken to represent interests of the disadvantaged people) over interests of politicians conforms to Chambers' (2005) view that participatory development requires principled leaders and change agents who ensure that disadvantaged community needs are not militated against where political elites or donors uncritically wield unnecessary pressure to have their interest considered in aid programmes. It is also quite informative about the role of politics in aid programmes that promote participation and partnerships between donors, politicians, civil societies, technocrats and disadvantaged communities. Political elites in the BEIP were invited to render 'political will' and to play advocacy roles to increase the acceptance and impact of the BEIP on the disadvantaged people within their political constituencies. Despite emerging as potential change agents, political elites used the BEIP as a campaign tool to entrench their personal and political interests. While politicians appeared to support schools that were within their political constituencies, the data show that they also wielded influence on the choice of schools where the BEIP was implemented:

> When we [technocrats] identified these schools and the same are presented to the politician … you know these are engaged in development of their area, you may find a politician dismissing a school or just striking a school [from the list]

> not because he is informed of its needs, but because he feels that, that area is not fully his/her supporter. So there are some vested interests. (Emmanuel)

The problem is that collaboration with political elites did not necessarily ensure that the schools that were most needy by virtue of the set criteria (or others) benefited from the BEIP. For example, in Western Province some of the schools that were dropped were considered more disadvantaged than those that were finally supported. As said before, the BEIP was withdrawn from all primary schools in Coast Province to pave the way for a World Bank (or USAID) project. Such withdrawal supports Ife's (2002) view that donor activities may disturb local processes of development without offering benefits of sustainable development (or change). The problem was that these communities had already undertaken needs-assessment and mobilized themselves in support of the BEIP. Such withdrawal reflected negatively on donor and government commitment towards enacting development cooperation and partnerships in ways that increase control of development by the disadvantaged people. The sector-wide approaches to planning policy change that led to the withdrawal of the BEIP meant that donors could implement projects where they chose as long as these projects fitted with the priorities identified in the Kenya Education Sector Support Programme.

The challenges show how cooperation floundered not just in the BEIP, but also in the broader policy within which it was seated. The potential and limits of aid assistance, technical expertise and representation to generate social networks through cooperation were apparent in the language of 'partnerships', 'development partners' and 'stakeholders'. Mainstream participatory development can use these terms to promote competition while also concealing power/control relationships. In support of this view, Sundukia, a parent/BOG member, said:

> The term *stakeholder* is more of a textbook thing, rather than the reality. When we come to the practice, it is completely different … To talk about stakeholder that levels the power relations. Each stakeholder has to believe that has a part to play. The other one is a *partner*. The other one is equally important. Starting right from the donors, we don't see that relationship of *partnership*. They will come and decide this is what we are giving and … for this [reason]. I mean, you can't call that partnership when one party is completely … making decisions on what is happening … and that whole thing flows in the whole chain. Basically, when you talk about stakeholders, I think it is more of a catchword in development. In practice, to a large extent there is nothing. We could have very few isolated cases where to some extent there is that stakeholder relationship. But from most of the [aid] projects in my experience, there is nothing like stakeholder. You [community] are simply seen as beneficiaries and you receive the little you can receive. Not only the issue of receiving [fund], even the information, everything ... I mean you are not taken as if you are in the same level right from the donor to intermediaries, implementing agencies to all that. Can you imagine the District Education Officer seeing himself as a stakeholder with a parent? Not until he

> unlearns, because even the way the District Education Officer will approach the community/parent is so paternalistic. It's just you are [up] there and I [parent] am [down] there, and they just take their positions. The moment there are such kinds of relationships, the issue of stakeholder becomes quite weak.

This subtle character of the discourse on partnerships to conceal power relations confirms Ife's (2002) view that pluralists' and elitists' discourses support cooperation, but also neutralize opposition to existing power relationships. In other words, they help legitimate the status quo. In the BEIP, the emphasis on achieving efficiency and cost-effectiveness of aid through technical expertise and politics of representation encouraged managerialism, rather than governance (Ife, 2002). Such managerialism (and related domination) was evident in the way partnerships in the BEIP encouraged upward rather than downward accountability.

Accountability

A core aim of participation and partnerships in the BEIP was to enhance transparency, performance-based management and accountability (GOK, 2005a, p. xii). Parsley (a technocrat) described accountability in the following ways:

> Accountability is broad … But, we [technocrats] have understood it in the straight way of efficient management of resources … In a nutshell spending resources for the purpose intended … and in an open manner.

This attempt to tie accountability to management was premised on the belief that participation of school communities through representation at the national, district and school levels increases accountability. As established in the management structures, the disadvantaged people could only participate through representatives. To achieve accountability and transparency, representatives in the SMCs/BOGs/PTAs were expected to be impartial, of high integrity and able to encourage competitive bidding in contracting and procurement of services and goods:

> Representatives are supposed to be people who are *impartial* … I mean having no vested interests. I would not like to have a situation where the owner of a hardware shop is the school chairman, so that most of the goods will be bought from his shop without competitive bidding. Too, we would like to have people who have *integrity* … I mean people who have high reputation in the community based on their character and conduct. (Emmanuel, emphasis added)

The emphasis on integrity encourages the view that aid power, technical expertise power and representation are not enough to enforce accountability and openness. That is to say, technocrats may have the technical know-how, donors may have the aid, and representatives may be democratically elected, but still struggle to be accountable and transparent in their development practice. Thus, integrity is a core measure of accountability and openness. To be considered accountable and transparent, change agents must demonstrate attitudes that aim to increase collective, not individual, benefits. These points are also set out in the data, for example Parsley described malpractices that backgrounded a GOK decision to enact structures to enforce accountability and transparency as follows:

> First and foremost, attitude must be touched. In Kenya and other African countries there has been ... hostility that came as a result of colonialism … that 'Mali ya Uma' [public property] attitude. If you recall, during the building of the railway, some communities were stealing some parts of the rail. You know because they considered it as not theirs as a community but of the foreigners … Therefore, because the foreigner is fixing it in their country, why can't they take it away from the foreigner? The foreigner also ... contributed by taking the African land. [This] was a way of telling people that they [foreigners] can also not be transparent … even when [the] country gained its self-rule, there was still that 'Mali ya Uma' business. People felt that they needed to take it [public resources] away. Also the government failure to harmonize or spread its benefits to everybody … people had this kind of attitude that since the government has been unfair to them, they also can be unfair to it. [Government failure to] enforce a harmonious … spread of wealth … equitabl[y] … either in terms of regions or salaries … has been a driving force for people to want to sabotage and benefit themselves. What I'm trying to say is this ... This [Mali ya uma attitude] is embedded in people's ways of life. Most people who are not accountable or transparent they do not do it because they are poor. They do it because they want to do it. They feel like … they are paying back in kind [punishing].

Technocrats saw accountability as involving politics, power, resources and responsibilities and where people chose to adhere to certain set standards. The need to effect transparency and accountability was informed by entrenched malpractices of the government, technocrats, donors and communities in previous aid interventions. The view that resources advanced by the donor through the government are for stealing (or sabotage) explains why technocrats adopted participatory and collaborative approaches to manage the BEIP. The problem is that the management structures the GOK used encouraged development 'managerialism' which tended to reify aid power and technical expertise, rather than 'governance' (Ife, 2002), which emphasizes people power.

Various strategies were employed to increase accountability. These included a phased process of implementation, which led the BEIP to be implemented in three phases (each spanning a period of three to five years), the introduction of a system of disbursing funds in instalments to match the revolving credit, and a system of quarterly reporting. For its part, the 'revolving fund' policy was meant to ensure that funds were released by the OPEC Fund to the GOK and then to the schools when needed. Technocrats considered the process of disbursement to be bureaucratic, as it stretched downwards from the supra (donor) through the Central Bank, Ministry of Finance, BEIP management structures at the central office of the Ministry of Education and district implementation committees to the schools at the grassroots level. To cut down on the risks of middle level institutions and commissions/interests accruing from the loan once disbursed, technocrats established the revolving fund. In practice, this meant that the funds could only be released to the school in instalments.

To enable the disbursement system to operate as desired, technocrats fixed the amount of money (or ceiling costs) each school was entitled to spend on the BEIP at 2.1 million Kenya Shillings ($US 26,923). The quarterly reporting system required SMCs/BOGs/PTAs to submit progress and expenditure reports of the BEIP to technocrats (Project Implementation Unit) and the donor through the district implementation units. The reporting system also required SMCs/BOGs/PTAs to post summaries of expenditures made on the BEIP on public notices and to maintain financial accounts and balance sheets.

In addition, a decentralized financial management system that allowed SMCs/BOGs/PTAs to undertake financial functions and responsibilities relating to the BEIP was also put in place. The system allowed schools to operate bank accounts through which the BEIP funds were directly credited. A key guideline on the financial management required SMCs/BOGs/PTAs to establish committees whose functions entailed contracting, tendering and procuring services-goods through competitive bidding processes. Technocrats' testimonies confirmed that decentralization of financial management to district implementation committees and SMCs/BOGs/PTAs increased accountability and transparency:

> … in fact that money is best accounted for at the district or school level than if it were in Jogoo (central office). The other day we were doing some of the reports on monitoring and the audit queries are now less than when everything was centrally controlled … the level of satisfaction … it is never 100%. There are always problems here and there … but they [SMC/BOGs/PTAs] are appreciating what is really happening to them and the sector. (Duncan)

The data indicate enhanced motivation and responsibility on the part of district implementation committees and SMCs/BOGs/PTAs. As a result of such motivation and responsibility parents invited technocrats to their schools to assess how they had managed the BEIP funds. Testimonies with technocrats indicated some degree of satisfaction with the way SMCs/BOGs/PTAs performed their financial management tasks. They also said that good management of the BEIP finances was evident in the quality of buildings.

However, there were issues relating to the monitoring and evaluation methods which were intended to increase prospects of continuity and commitment to sustain the BEIP on the part of disadvantaged people. To ensure that financial management rules and regulations were adhered to and to increase responsiveness of the BEIP to its objectives, a joint monitoring and evaluation system, which allowed the donor, technocrats and technical experts to collectively undertake assessment of the progress of the BEIP once a year for the duration of the project cycle, was put in place. Joint evaluations were meant to complement the monitoring and evaluation activities that district implementation committees undertook in each school fortnightly and which SMCs/BOGs/PTAs and parents undertook continuously on a daily basis. Although the policy on monitoring and evaluation gave:

> the Ministry of Public Works and Housing … mandate to inspect the project, as they would at any other time... they will not come to the site unless they are instructed to do so … By whom? That one now remains [an issue] … Is it the client who tells them to come or should they be able to judge within the progress of the project that this time round I need to be on site to see what is happening. (Benjamin)

Such ambiguity in monitoring and evaluation was perceived to perpetuate a 'project money' attitude on the part of technical experts, some of whom demanded to be paid *kitu kidogo* (bribe) by SMCs/BOGs/PTAs to visit the project sites even when the GOK had paid for such operational activities. Such unethical practices represent what would be seen as a loophole created in the funding and information systems of the BEIP to promote control and domination by technocrats, technical experts and the donor and to deny SMCs/BOGs/PTAs members payment for services they offered. In support of this view, Jasmine, a member of a district education implementation unit, said:

> … these are some of the strings attached. You cannot be given this money like that. There has to be somebody to benefit. Employment is provided for their … people and those people are expatriates, they are the ones who get more in payment. You know people like [grassroots representatives] are told … you are project vocal points. And then you are … told you people … are government employees … You are not supposed to benefit … You are donated by your government towards this project to assist … Yet, this person has been donated by his government, but, them a lot of money, us, nothing … One time, I was told, 'you are sacrificing for your community. You don't need to be paid'. So, those are some of the strings attached. The donor has to pay 'watu wake' [his people] … and then the local person who is supposed to be the implementer of the project at the district or the community level [is] supposed to give sacrificial services.

By only financing the monitoring and evaluation activities performed by the donor, technocrats and engineers, the GOK devalued the monitoring and supervisory activities of SMCs/BOGs/PTAs. Again, the schools incurred running costs through printing and purchase of material for structural designs which initial BEIP budgets did not anticipate. When school heads requested reimbursement for such costs, they were said to be corrupt. The data in which these undesired outcomes were noted also pointed to ulterior motives where technical experts failed to fulfil their obligation of designing and drawing bills of quantities even after SMCs/BOGs/PTAs invited them. Again, when the technical experts finally visited the school, they would disapprove of some of the decisions and activities SMCs/BOGs/PTAs undertook in their absence.

While acknowledging the worth of frequent monitoring by technical experts, from a social justice perspective (Ife, 2002), the attempt by technocrats to enact structures that increased control by technical experts contravened the BEIP aim of enabling SMCs/BOGs/PTAs and the disadvantaged people to enact these rights. In the perspectives of SMCs/BOGs/PTAs the practices of participation and collaboration relating to monitoring and evaluation entrenched corruption, not accountability and transparency. Indeed, the methods of accountability represented technocrats' and donor attempts to control disadvantaged people's development processes without offering permanent solutions to the root causes of their poor conditions.

The SMCs'/BOGs'/PTAs' testimonies showed that in joint monitoring and evaluation activities, donors reportedly accessed higher remunerations compared to their local counterparts. In this respect, the BEIP barely enhanced equity. Again, the OPEC, for example, required the GOK to operate an off-shore bank account that was perceived to generate interests from the BEIP loan but whose books of accounts disadvantaged people could not access. Thus, donors were unaccountable where they endorsed conditions to support power and market shifts from the disadvantaged people. Technocrats were also unaccountable. Mapatano, a political councillor, said:

> We were not accountable. Some donors are not accountable. They [donors] say you can only get this money if you employ the executive director coming from country X and you will pay this and this and discuss this and that ... they earn a quarter of that money back to their country. This is [un]accountability ... and then you [citizen] pay both the capital and the financial aid.

This view that both donors and technocrats are not accountable underscores the limitations of upward accountability. The problem is that donors and technocrats do not see the need to account to aid recipients, because in the words of Reuben, a head teacher:

> First and foremost the donation is given because you are not self-sustaining. That is why you need that assistance. It does not mean that when you will be given that assistance … you will be able to support yourselves in other areas of need ... Even in the same area you have been assisted. But you have not been able to be independent in your undertaking, at present and in the future. Now the way you take care of the resources that have been extended to you, will be able to determine what other assistance might be presented to you. Therefore, that should be the principal focus, that when you take care of what you have

> been given, how you use it, in a transparent manner and at the end of it, you can be able to draw a balance sheet and say income is equal to expenditure and anybody can see, you expect to score the highest mark possible and get confident that the hand [donor] will be able to come back again and possibly assist you. If you take care of something small you will also be entrusted with a bigger responsibility. That is the way I see it. And it [accountability] should be seen … from all dimensions … donors, politicians, peers, mates and those who are below should be able to see … what has been achieved through this kind of funding/assistance.

SMCs/BOGs/PTAs have come to believe that they need to create confidence with donors through reports and balance sheets so as to attract further aid. They are also aware that aid enhances the vulnerability of disadvantaged people because it does not offer permanent solutions in the present or the future. It is, thus, fitting to argue that upward accountability legitimates the status quo; and encourages power and market shifts to the already powerful, and, partnerships within an aid delivery system are unsustainable because they facilitate dependency on donors.

While upward accountability to the sources of funds is critical (Craig and Porter, 1997, 2003), it can be questioned to the extent that this accountability retains control to the already powerful. This way, partnerships that encourage accounting through balance sheets (reports to the donor and displays in public notices), though undeniably critical, can be seen to redefine openness in ways that negate donor-to-government answerability (or downward accountability) to the people.

The methods used to enhance collective responsibility to the disadvantaged people included the aid agreement and performance-based contracts that Permanent Secretaries of the involved Ministries signed. SMCs/BOGs/PTAs, on the other hand, were expected to post expenditure summaries of the BEIP fund on public notices. Though meant to enable technocrats and the donor to be more accountable to the people, these methods actually negated government responsibility and downward answerability, as one educationist, Mapya, reported:

> Performance contract … is one way of trying to achieve that accountability. Assuming that all things are put in force … we should have benchmarks against which to gauge our performance … the government has put in place these performance contracts with an aim of improving services to the people. This means that when officers do not perform their duties and tasks in time, the citizen has a right to complain and then I [officer] will be held responsible.

The aim of performance contracts is to enable accounting and authorized officers (Permanent Secretaries, Chief Executive Officers) and heads of departments (who are policy-makers) to achieve set targets of delivering quality services to the citizenry and other service users. The performance contract is, thus, a tool for measuring the quality, quantity, efficiency and cost-effectiveness of government services to its citizenry against predetermined standards such as those relating to the earlier discussed methods of enhancing accountability. The performance contract is also claimed to enable service users to hold governments accountable.

Despite this political appeal, the problem is that performance contracts neither increased accountability to the people, nor offered a legal mechanism to provide citizens with an effective mechanism to voice their concerns. To the question who hears the voices and complaints of the citizenry, Mapya avowed:

> That is now the problem ... The government ... is amorphous, but the performance contract will enable my supervisors such as the Permanent Secretary [who is] the accounting officer for the Ministry to hold me accountable. Before he accounts [to the Executive] he will find out who has messed me? (Mapya)

The bureaucratic structure within which the BEIP was implemented also meant that Permanent Secretaries signed performance contracts on behalf of the Ministry of Education. A Permanent Secretary does not sign this contract as an individual but on behalf of "the whole body" (Mapya). The question is where officers do not achieve set targets, who has failed? Is it the Permanent Secretary or the 'amorphous government'? The answer is as follows:

> You [Permanent Secretary] are signing that [contract] given the responsibilities and duties that are charged by your office [function]. You need to ensure that what you are signing for is being accomplished. Otherwise why are you signing? Like if I am supervising three people and I have been given a deadline, it is upon me to make sure that I have the [resources, tools]. [Not] making necessary channels to do my work, then I am failing ... The point is, it is not the person. We are talking about the office. Offices are charged with certain responsibilities which are supposed to be implementing government projects/programmes. (Mapya)

As noted, performance contracts do not commit duty-bearers. They commit the office (functions and positions). Performance contracts can, thus, be seen to negate government responsibility to the people while encouraging development managerialism (as opposed to governance) and corporatism (as opposed to cooperation) through practices of partnerships and participation. According to Ife (2002), development managerialism and corporatism denote third way practices in which due to the failure of the first (government) and second (private) systems, Third World governments are encouraged to copy corporate ways of thinking and doing business to increase efficiency and cost-effectiveness in human development. For example, a member of a BOG argued that policies and methods which encouraged the BEIP to be implemented in three phases and the fund to be disbursed in instalments were responsible for "many uncompleted projects which had been donor funded". This occurred because once a new government comes to power, donors withdraw funding on claims of poor governance when their "contact" persons are not re-elected or re-appointed (Amani). The methods of encouraging downward accountability floundered to effect the desired outcomes because they were designed to satisfy donor and technocrats' personal interests. In pursuit of such interests, the practices of participation and collaboration relating to aid agreement and performance contracts not only privatized decisions and resources (rolled back state regulatory role and its legitimacy), but also further marginalized the disadvantaged people and institutionalized poverty as Sundukia (a parent/BOG member) asserted:

> In a country where development is linked to personalities other than clear policies, then automatically that means that everybody who has been excluded from education cannot climb up to the top, cannot influence policies, and the moment you cannot influence policies, that means exclusion and once you are excluded then poverty is with you. And it is not only with you for days, it can even be institutionalised. This is because if you are poor and you cannot take your child to school, you bring up another poor person [and the chain continues]. With time you realize, regions, communities, [and] families are completely excluded.

The data reported that politicians were more able to enforce downward accountability because they are not only democratically elected but also because their main role is to ensure access to rights by the citizenry. However, their behaviours in the BEIP partnerships compromised this role. Emmanuel argued that when political elites and civic leaders turn aid projects into campaign tools:

> ... the project ceases being seen as a responsibility that the government is playing but now the Member of Parliament wants to capitalise on the project for his own gains. In fact that one might affect participation because there are some people who may feel that *hii ni* [this is a] project *ya mjumbe* [of the Member of Parliament] so we won't [participate] ... (Emmanuel)

The redefinition of the BEIP into a campaign tool represented a process of individualizing, as opposed to generalizing (or spreading) benefits to the community. This privileging of personal over collective needs arose from the formation of partnerships through principles of competition and representation. By engaging politicians in the BEIP partnerships, the GOK satisfied the alternative development view that political will and empowerment cannot be achieved outside of government systems (Friedmann, 1992; Ife, 2002). Nonetheless, the behaviours of political elites to confirm Brown's (2004) view distorted authentic political relationships by using the BEIP to legitimate and entrench personal political interests onto their constituents. For example, the use of the BEIP as a campaign tool had a double effect on participation of the disadvantaged communities. First, it compromised downward accountability by discouraging participation of electorates who did not identify with ideologies of the elected political elites. Secondly, the relational dynamics that emerged encouraged communities to perceive political elites as 'donors'.

Such redefinitions led political elites to make false promises to their constituents to justify their statuses as 'political leaders', 'aid assistants', 'equal partners' and 'representatives' of disadvantaged people in the BEIP. As noted in the data, most SMCs/BOGs/PTAs contended that by funding the BEIP wholly, OPEC was saving communities from domination by political elites. Paradoxically, by arguing that communities should not be asked to contribute monies is to support power and market shifts to the donor and agree to bureaucratic power and dependency. On these views, head teachers argued that:

> … the conditions which are laid [out] are really limiting us to implement the project. When money comes … you are told that you have been allocated 2 million to build a laboratory. The community is looking forward to see that project completed … Now we have been given half a million, only months have gone, it is only papers. Every time papers and then you cannot move. I think the guidelines, the bureaucracy which is there is really too strict. It slows down the implementation of the project … The government procurement procedures should be restructured. (Reuben)

> There is also the colonial status of the country … You find today you procure... materials at a particular cost. In a few days or weeks … you find that the prices shoot up. Now you have to reschedule everything because the supplier will come down running … telling you, mwalimu [teacher] we cannot supply you with these, because we cannot do it at this price. That also hinders our participation. (Benjamin)

The linkage between colonialism, bureaucracy and markets confirms that due to internationalization of capitalism, commodity (and service) prices and decisions are made at levels that are beyond the proximity of governments (Ife, 2002). The validity of partnerships aimed at increasing accountability through contracting, competitive bidding and performance contracts can thus be questioned on technical grounds. Head teachers resorted to goodwill and personal negotiation skills to establish networks with suppliers of markets and goods, not GOK circulars. To the extent that head teachers depended more on cooperation, rather than competition, the GOK commitment to enact structures and partnerships that increase control to the disadvantaged people can also be questioned on moral grounds. The methods of accountability encouraged the view that donors and governments are self-reflective and self-regulatory (Chambers, 1997). Yet, partnership practices taught parents that neither the donor nor the government can be held accountable to the people. For this reason:

> What I can say in terms of … these international donors and their partnerships with the government is that, they are subjecting us to corruption. That is the best

> thing I can say... You cannot give me a project of 2.1 million and give me five hundred thousand and let me hang there for two to three years, as if nothing is happening there. I will be forced to pick from here and there so that the project can go on. So they are subjecting us to corruption. (Mapatano)

For example, the education sector lacks "the capacity … attitudes... the will to change things, to want to focus on change as opposed to structures. The moment that is wanting then we cannot talk about downward accountability" (Sundukia). This distinction between structural and change reforms as detailed in later chapters is critical to understanding empowerment, transformation and sustainable development. As critics of modernization (Isbister, 1991) have argued, the structural approach used to promote accountability focused on creating partnerships to solve the poverty and rights challenges of disadvantaged people but cemented into neoliberalism. By focusing on structural, rather than change-oriented reforms, the Ministry of Education was perceived to lack in vision and responsiveness to the needs of disadvantaged people. To be responsible, the education sector needed to enact structures that balanced between structural and change reforms. Core strengths of the BEIP came from the attempt to provide structures and opportunities for GOK to promote development cooperation and partnerships with donors, civil society and disadvantaged people. Cooperation, partnerships and accountability represent possibilities for change.

Conclusion

This chapter has shown that despite the GOK's professed desire to increase the power of disadvantaged people over their futures, the attempt to create partnerships that aimed to maximize aid benefits actually increased power of elites. The use of technical expertise, aid assistance and representation negated equal partnerships, encouraged competition rather than cooperation, and upward rather than downward accountability. Thus, the BEIP partnerships risked relegating GOK responsibility to the disadvantaged people while encouraging dependency on aid.

Notwithstanding such shortcomings, the GOK attempt to connect disadvantaged people with institutions that governed their lives points to possibilities for change. SMCs/BOGs/PTAs confirmed the empowerment claims of partnerships by questioning why technical experts spent tax-payers' money to hire cars and their failure to perform roles as expected when they continued to receive salaries. This highlights a significant convergence of participatory development in aid projects with state-led development. Paradigmatically, it underscores the point that disadvantaged people can hold technocrats (or government) accountable and responsible to them on the basis of their rights of self-determination (participation), association (partnerships) and citizenship. Since donors and governments are not self-reflective, self-correcting and self-responsive as assumed in upward accountability, an empowering and transformative participatory development must promote participatory practices, government partnerships with civil societies and local communities through structures and discourses of rights and citizenship. This will enable disadvantaged people to claim accountability and enforce government responsibility based on their human and civil rights. The next chapter examines the extent to which participation satisfied principles of change from below.

Chapter 5

The Methods, Process and Outcomes of Participation in BEIP

Introduction

The previous chapter showed that an emphasis on structural, rather than transformative reforms meant that the BEIP partnerships accentuated vulnerability of disadvantaged people. This chapter builds on this view by exploring how SMCs/BOGs/PTAs participated in the BEIP and the impact of that participation. It critically examines the structures and approaches that technocrats used to facilitate participation in the BEIP, to ascertain the extent to which it satisfied principles of change from below. It then argues that although the BEIP provided structures and opportunities to increase participation, the structural approach technocrats adopted obscured the anticipated empowerment and social change benefits of the *participatory process*. The three sections covered are planning, implementation and monitoring, and evaluation. In each section, focus is upon objectives, methods, meanings and outcomes, and the opportunities and challenges these features posed to the actualization of 'full' participation in development by the disadvantaged people whose lives the BEIP was meant to improve.

As noted in previous chapters, development practice and analysis should assume a broad systemic perspective (holism) in understanding particular problems or processes (Ife, 2002). The approaches adopted in the BEIP were designed to promote participation as a process, not just as separate means to ends. This approach is supported by Ife, who promotes the view that the process and outcomes of development are inseparable, because means and ends are morally connected. To protect the integrity of the process, means should be treated as outcomes. Likewise the means and ends of participation are inseparable. Means can become ends and ends can become means, especially because participation is both a democratic right and a critical learning process (Burkey, 1993; Chambers, 1997). These relationships require change agents to act in ways that do not disenfranchise the disadvantaged people.

Building on the means, ends and process principles set out in Chapter 3, the data presented in this chapter show that despite policy claims to holism, in practice, technocrats separated the means from the ends of participation. This separation meant that participation was prescribed as a technical panacea to the rights and structural challenges school communities experienced. This emphasis on structural outcomes, as opposed to rights (where the process of participation constitutes the right of self-determination) compromised integrity of the participatory process.

The next section shows how such separation reinforced top-down approaches to change while limiting SMCs/BOGs/PTAs and disadvantaged people's participation in key decision-making processes.

Project Planning

As the broad literature on participatory development suggests (Cornwall, 2004; Ife, 2002), participation in planning aims to make policies and development sensitive to community needs (GOK, 2003b, 2005a). According to official documents (GOK, 2002b, 2003b), the GOK enacted structures that were designed to achieve this. Planning entailed analyzing needs, capabilities, resources and activities of redress, including ways to mobilize extra resources to cover funding deficits. The extent to which school communities participated in project planning can be taken as an indication of their level of control of decisions within the BEIP. To understand the degree of such control, the next sub-sections explore the role school communities played in identifying and prioritizing the BEIP objectives.

Stating Objectives

Given the focus on empowerment and social change, it could reasonably be expected that school communities would have played a key role in establishing the objectives and approaches adopted in the BEIP. However, documents (GOK, 2002b, 2003b) suggest that technocrats, in collaboration with UNESCO, determined the objectives and the approaches adopted. Further, evidence suggests that, at the time of determining objectives (and approaches), the project proposal (the development plan) was primarily a fundraising tool, rather than a means to increase participation. Thus, because of the need to meet the requirements of donors, technocrats established objectives which *they* perceived would address structural and rights/needs of school communities. On this view, Bushie (an educationist) stated:

> Donors come … at the national level. Negotiations, conclusions and signing [of the agreement] is done here. So these things start rolling once signatories have been attended and terms have been agreed upon. So … we *tell* stakeholders *down* there (school), that you are stakeholders, you are parents, you are expected to do this and that. *We are not asking them, what are your ideas*? Can we incorporate your ideas in this particular project called OPEC? You are already telling them that this is your responsibility … do number one, two and three. (emphasis added)

This 'planner-centered' approach (Michener, 1998) to participation contradicts principles of change from below. Change from below necessitates that communities initiate development either by themselves or with the facilitation of change agents (Ife, 2002). The use of words such as 'telling' and 'down' by technocrats in reference to school communities respectively denote undemocratic and top-down mindsets and related approaches to development. These mindsets are confirmed in the data provided by Antoinette who pointed out that participation in project planning occurred through representatives. She pointed out that:

> [T]he planning level … is at a point where there was involvement of key … Semi Autonomous Government Agencies ... We had involvement of Kenya Institute of Education [and] different departments in the Ministry of Education … so that … the donor would come into an agreement with something that has been … accepted by the stakeholders of the headquarter levels before we went down to the consumer … What I experienced first in the design … of the programme itself – it was a brain storming [exercise] with different Semi Autonomous Government Agencies … so that we would know how to *trickle down* the whole project, down to the consumer. (emphasis added)

This confirms that technocrats are involved in processes of policy translation and that their decisions are not value free (Ife, 2002). According to the participatory development management literature (Chambers, 1997), an aim of participation in planning is to bring decisions closer to the subjects of development and, thus, enable these subjects to exercise their right of participation. The statement of project objectives by technocrats at the outset meant that neither the SMCs/BOGs/PTAs nor disadvantaged communities actively participated in constructing the BEIP. According to Bushie, these individuals and groups were excluded from project planning because the project " … is still within the management structure of the central office. It has not rolled to the district, divisions, zones … and it is … the way government programmes and projects are run".

This idea that project planning is a responsibility of the national office denied the disadvantaged people and their representatives their democratic right of agency and self-determination in setting the BEIP objectives (or determining the Vision). Relating to their meanings of participation SMCs/BOGs/PTAs testified that:

> … participation is … being involved in the identification of the project and in this case, in OPEC, it is not the case because they [technocrats, donor] already know what they want to do … actually the [priority] areas have been undertaken … we should [also] be involved in … evaluation of the project, in the initial planning, even sustenance … awareness creation on how to implement the project. (Geoffrey)

As noted, SMCs/BOGs/PTAs desired to actively participate not just in setting objectives but also in all BEIP processes. The aim of such participation (which was also set out in the poverty reduction strategy paper (GOK, 2002b)) was to elicit commitment and ownership of decisions and the accompanying processes of implementation. For example, a key decision made during project planning committed school communities in their absence to implement the project. This meant that 'participation' was predetermined. This failure by technocrats to involve school communities in these initial decisions meant that the project and its participatory processes were imposed from above.

The next section offers further insight into the nature of participation in planning by arguing that although technocrats created structures and opportunities for SMCs/BOGs/PTAs to (re)define the needs addressed in the BEIP, their participation was tokenistic-coerced or passive-instrumental.

Statement of Needs

Data indicated that technocrats appreciated the need for SMCs/BOGs/PTAs to take part in assessing the needs of their respective schools as a way of empowering these communities to govern and manage the BEIP more effectively. For this reason, technocrats enacted the needs assessment policy. Accordingly, technocrats professed a desire to allow participants to determine their development needs. This is supported by Antoinette, who reflecting on the BEIP benefits, said:

> … the needs assessment … is an opportunity for the school community to participate … that is one because we sought their opinion. Another one is the commitment that finally the project will be given to the community, so that the community can gain from it.

By offering SMCs/BOGs/PTAs space to (re)define their needs, the BEIP cemented into the idea that any development intervention must increase disadvantaged people's "power over statement of needs" (Ife, 2002, p. 57). Nevertheless, the way technocrats facilitated SMCs/BOGs/PTAs to assess and (re)define their needs meant that, in the main, participation was coerced. Official documents (GOK, 2003a, 2003b) show that the needs assessment aimed to affect participation and collaboration through a series of prescribed activities. Firstly, the approach adopted by technocrats was to develop needs assessment tools and to test their efficacy with sample schools before these could be used in all the targeted schools. The "pilot testing in five schools" (GOK, 2003a, p. 11) aimed to establish efficacy of the questionnaires technocrats used to assess schools' needs before these tools were applied to the rest of the 350 schools that took part in the BEIP. The pilot study was also meant to address ambiguities in the questionnaires and harmonize areas of need identified therein with those of the school communities. Though understandable from an elitist viewpoint, SMCs/BOGs/PTAs described participation as 'extractionist' because they felt cheated out of their rights, knowledge, time and resources by participating in the pilot study. One parent, Oromosa expressed this sentiment as follows:

> ... so we are seeing that we are being used as experiential animals ... the project ... is a question [of] experimenting it with us. Whether ... successful or a failure, it will be implemented in other areas, not where it is piloted. So it is exploitive in nature. Though the exploitation is silent, it looks exploitive.

As suggested earlier, the failure to implement the BEIP in some of the schools where the pilot study was undertaken, together with the withdrawal of the BEIP from primary schools in Coast Province, led SMCs/BOGs/PTAs to believe that technocrats were experimenting with untested strategies on the disadvantaged people without addressing the root causes of their poverty. As the broad literature on participatory development (Shivji, 1999, 2003) attests, a key problem is that the questionnaires used could neither determine with ease the quality of participation nor assess the levels of poverty and micro, meso and macro level inequalities (Pieterse, 2002).

The potentials and limits of these questionnaires can be better understood by exploring the other two activities: statement of needs and consensus-building. These features ran simultaneously in practice but for ease of reference the data are presented separately. The needs assessment processes, ideally, aimed to enable communities to determine their needs and build consensus in decisions. It also aimed:

> to assess the degree of [communities' interest in the project]; develop interest and seek possible participation in the processes; to discover and understand alternative ways of defining goals and objective of the project; to bring to light information about the programme/people; to begin identifying existing projects, assessing experience and determine institutional resources available so as to determine what can be drawn upon and what needs to be added from external sources. (GOK, 2003a, pp. 12–13)

Empirical data supported these aims where Antoinette reported:

> Needs assessment was one of the exercises that was very key ... We were trying to assess the priority needs of each school so that we will be sure of what they want us to do ... If it has to be participation, it is not about the ministry dictating what they want to do in the schools. It was about the schools being involved in deciding their priority areas of needs... I would say that the needs-assessment was an eye opener to enable the consumer to accept the project and move it.

This need to secure community commitment and support for the BEIP necessitated the planning team and project coordination unit to enact a series of methods and processes that culminated with communities (re)defining their needs and also attaining consensus-building on the decisions that emerged. To (re)define needs SMCs/BOGs/PTAs used the said predetermined questionnaire. The technocrat designed questionnaires were based on the BEIP objectives: school management and governance, physical facilities and other teaching and learning resources, wellbeing (healthcare, sanitation) and participation and community (mobilization). This meant that the identification of needs began with determination of the project objectives and selection of schools by the technocrats. Moreover, rather than drawing information directly from communities, technocrats initially used secondary information to select disadvantaged communities and demarcate their needs (GOK, 2003b).

Further, the data show that district implementation committees and political elites facilitated these selection processes. The problem with using secondary data (or third parties) to establish 'needs', as revealed here, is that paradigmatically, the right of participation is treated as an information gathering event as opposed to a democratic and learning process (Burkey, 1993; Chambers, 1997; Ife, 2002). Accordingly, the tools technocrats used to assess needs and the related processes arguably masked democratic and learning benefits to SMCs/BOGs/PTAs and the disadvantaged people these groups were meant to represent. This is because these communities played no role in determining such tools and methods of analysis. Technocrats administered two sets of questionnaires. The first prescribed head teachers' participation, which Kagendo, a head teacher, described as nothing more than "completing the questionnaire". The second defined the participation of educationists, planners, members of SMCs/BOGs/PTAs, area education officers or inspectors of schools and village-community leaders. Documents show that technocrats held three-day meetings with SMCs/BOGs/PTAs to determine needs based on this questionnaire. As detailed under consensus-building, the statement of needs was an iterative process where these groups identified and ranked needs and actions of redress based on the (un)available resources (human, social, economic, and political).

According to technocrats' testimonies, the questionnaires were comprehensive. They covered most information necessary for technical planning and resource allocation. According to Ife (2002), such expertise is necessary for information gathering and data analysis. While acknowledging the value of inferred knowledge, the paradox is that, by using questionnaires, technocrats appeared to reduce communities into objects – passive recipients of predetermined development plans, needs and decisions. This denied disadvantaged people the rights of choice, agency and decision-making and, as detailed later, privileged technical over (or devalued) local knowledge.

According to Ife (2002), communities must determine their own needs and tools of analysis to avoid imposition of needs and methods determined elsewhere. Empirical data recorded awareness on the part of technocrats to avoid dictating needs to the disadvantaged people. Describing BEIP outcomes, Antoinette said:

> My experience … was that there was a lot of acceptance of the project by the SMCs, teachers, District Education Officers. Why was there so much acceptance? It is because at one point we are giving back the project to them and no longer calling it the ministry of education headquarters project, but calling

> it the schools' project. We referred to it as 'your project'. Though we were the designers, though the money was being facilitated through the headquarters, we returned the whole project to them for the purpose of participation and ownership …

Ife (2002) contends that communities should own the project from its *inception* and control the *whole* process of development, including the design stages. The data show that only SMCs/BOGs/PTAs took part in the (re)design stage: the broader community of disadvantaged people did not. Upon designing the intervention, the technocrats 'returned' the project to these communities. Nevertheless, the SMCs/BOGs/PTAs and the disadvantaged groups they represented accepted the BEIP, but their participation was replete with challenges.

Their participation in planning processes was only a reaction to needs and decisions predetermined by technocrats. Given that the communities did not determine their own needs, methods and solutions, under the BEIP, participation was both coerced and disempowering. Sundukia, expressed it as follows:

> You cannot tell people to participate when you have already drawn the playing field [project, questionnaire, needs, and decisions] and then you tell them come and play in the field I have already put. You are not sure whether some would have wanted to play outside. But you already tell them you guys come I want you to play here … And now you are equal partners, start participating. That coerced kind of participation, not many people … really want to participate. People mainly want to participate in issues they have made decisions about … themselves.

There is ample evidence in technocrats' accounts that participation in the needs assessment assumed consultation, telling or informing. Such participation is overly nominal (Cornwall, 2000) and tokenistic (Arnstein, 1971; Ife, 2002). It can, thus, be questioned to what extent the process of assessing needs undertaken in the BEIP increased disadvantaged communities' control and power over determination of needs, ideas and life chances. The idea of participation, through representation, meant that disadvantaged people themselves did not participate in the needs assessment meetings. The needs assessment structures and the practices these generated, thus, excluded disadvantaged people from enacting their right of participation and obscured benefits of learning (empowerment and social change) which they would have had. Indeed, the attempt to use generalized tools to determine needs constrained the alternative choices of school communities and, as said before, negated contextual differences and 'universalized' development of the disadvantaged people by precluding needs that fell outside the provided alternatives. Again, most SMC members with low levels of literacy least understood the second set of questionnaires and the method of analysis.

As detailed later under the consensus-building section, it is for this reason that teachers played a significant role in completing the questionnaires on behalf of the disadvantaged communities and interpreting these for SMCs/BOGs/PTAs. Although technocrats described participation as a right and argued that school communities 'fully participated', their attempt to determine needs and dictate resources meant that development under the BEIP was conferred downwards by the state (or donor) – development was something done to communities not something communities originated. In other words, under the BEIP, participation in development remained "a discourse of the powerful about the powerless" and a dictatorship of needs and resources (Ife, 2002, p. 67). The practice of participation in the needs assessment and consensus-building limited agency by retaining decision-making authority with technocrats and excluding disadvantaged people from enacting these rights. Sundukia expressed these views as follows:

> Of course … participation to them [technocrats, donor] means identifying needs … A project is supposed [to] help people identify their needs and go further than that. But now if participation is used to make people identify needs, really it is so limiting, because if I identify my needs and you take them, whatever solution you give is according to you, whatever resources you attach is also according to you. That is very much limited … When we are talking about stakeholders' participation, we are talking about *power sharing* … all we know is that power is very sticky … power determines resource sharing ... very few people [donor, technocrats] are really in the bottom of their heart committed to share power … and … resources … mainly the participation is limited to processes that enable the communities to identify their needs and plan … but *participation involves control* [of] resources and money … when you want to make a decision to buy this and that they don't want their beneficiaries to know that … So their participation is still very much naked. [emphasis added]

This way, technocrats risked using the coopting language of 'representation', 'questionnaires', 'stakeholders', 'full participation', 'beneficiaries' and 'needs assessment' to silence dissenting voices (and with it the kind of participation they had hoped for) and legitimate the status quo. Bushie said:

> In fact these things [projects] are rolled down, already with specifics, done [and] defined role of each stakeholder. So you end up being told, here is a programme and this is what is expected out of you … So *you are trying to align yourself to fit within the system of what you are being told to do*. [emphasis added]

Such depoliticization of the right of participation arose because of the need to achieve consensus-building through questionnaires and representatives.

Consensus-building

A belief in consensus-building led technocrats to create opportunities for SMCs/BOGs/PTAs to participate in decision-making forums in the BEIP. Consensus-building was meant to secure ownership and community commitment to sustain the changes the BEIP effected to the disadvantaged communities. In tandem with Ife (2002), consensus-building forums were expected to increase SMCs'/BOGs'/PTAs' and disadvantaged communities' control over their participation. Consensus was expected to emerge through the process of the needs assessment described above. Paradoxically, despite a rigorous process of assessing needs, the use of questionnaires and representatives, as said before, meant that the 'participation' was imposed on SMCs/BOGs/PTAs and the disadvantaged people from above.

Notwithstanding the significance of ownership and sustainability, it is one thing for communities to participate in implementing predetermined decisions and plans, it is another to claim ownership and be able to sustain a project, such as the BEIP. Proper ownership entails 'control' of decisions, resources and processes (Ife, 2002). Although such ownership was hardly actualized in the BEIP, it can be argued that participation resided with the people. This is understandable when the structural and rights approaches that technocrats assumed are juxtaposed with the type of schools the project supported.

As alluded to earlier, most schools had been built by communities through harambee processes. Again, the selection conditions demanded that participating schools must be "initiated and managed by the community [which must demonstrate] awareness of its needs" (GOK, 2003b, p. 16). Participation of these communities was, thus, a matter of principle. Since communities neither initiated the project, nor made critical decisions, whose development were they to participate in and own? This is not just a question of ownership and accountability, it is also one of agency and efficacy of participation to effect sustainable changes. Despite claims to ownership, the idea of deciding for disadvantaged communities and giving them aid resources to own, as detailed in later chapters, can be challenged on the basis that aid development is unsustainable.

The data with technocrats demonstrate awareness that participation is a process through which school communities exercised their democratic right. However, those who participated in the consensus-building meetings in the BEIP were not the real decision-makers. Consensus-building meetings are public arenas in which the disadvantaged participated. However, under the BEIP, the real decisions remained made in private forums that were inaccessible to those at the grassroots. Consensus-building through SMCs/BOGs/PTAs and technocrats meant that these meetings were only attended by representatives. Moreover, the use of questionnaires meant that SMCs/BOGs/PTAs built consensus on predetermined decisions and areas of need. A member of a district committee expressed this point as follows:

> If you want to attend a meeting and you want to get involved in that discussion, I think that is participation. The meeting has its objective and purpose. If it is a planning ... or consultative meeting ... most of our people [communities] are not even consulted in the projects they see around ... they are never consulted. Projects just come and then they are asked, to be involved and own the project. This is what people usually come and tell [us] … most of the time we feel that most of these projects are being forced down the throats of our people, because in the initial stage, the initiation process, they are not consulted on the projects. So somebody comes with a project and tells you ... you don't have facilities, he will give you good facilities. [Nashika)

Such participation also risked legitimating the status quo. According to Ife (2002), different types of participation (including representation) should be encouraged to increase inclusiveness. However, as said before, the use of representatives, to reduce multiple viewpoints, meant that the disadvantaged people were excluded from key decisions. Ultimately, either due to lack of time, or in the name of 'consensus-building', such decisions were rushed to satisfy donor and technocrats' interests. The reference of 'statement of needs', as 'assessment', or 'verification' suggests that technocrats used needs-assessment less to build consensus but more to justify predetermined decisions and actions as a technocrat avowed:

> … [F]or the first time we [technocrats] did not want to be told that we are writing a document from Jogoo house, for Jogoo house [by] the people [communities] … [T]hat's why we involved all stakeholders … [T]he process was participatory. (Duncan)

Though meant to ensure commitment and participation of SMCs/BOGs/PTAs and disadvantaged parents, such commitment was lacking, partly because "this was a project that was just coming in … The level of participation of the parents … was [limited] to be[ing] told this has happened ... [T]here … was no statement of agreement that you will contribute, that brings in the hitch ..." (Emmanuel). This point that the failure to sign an agreement with the disadvantaged people relegated their participation confirms Ife's (2002) view that stating needs is like 'writing a will' – it entails decisions, commitments and responsibilities.

Relegation of commitment and benefits of ownership also arose from technocrats' views that school communities needed to participate, but not contribute monies because they were poor. Paradoxically, by assuming that they were funding the BEIP wholly through aid, technocrats devalued local knowledge and potential. It is, thus, not perchance that disadvantaged people hardly participated in decisions and SMCs/BOGs/PTAs found it difficult to mobilize additional resources from parents.

> It is only SMC members who come to sit and deliberate on issues pertaining to the project. Otherwise, community participation as such … because the donation

> is fully funded … there is no way the community will come and say we are participating by providing funds here … At the moment no. (Chamkwezi)

Considering that participation has normative and technical values (Ife, 2002), the practice of participation in planning processes suggests that technocrats enacted and implemented participation as a technical panacea to the structural and rights challenges the disadvantaged people faced. The next sections argue that the structural approach to participation redefined consciousness-raising and community-resources mobilization as 'sites' for (de)valuing local potential, despite technocrats and SMCs'/BOGs'/PTAs' awareness that these features were critical empowerment and transformative opportunities to disadvantaged people.

Consciousness-raising and Mobilization of Resources

Consciousness-raising

The methods selected by technocrats to effect consciousness-raising were advocacy and capacity-building (GOK, 2002b, 2003b). Through these methods, the technocrats aimed to 'sensitize' the disadvantaged and the SMCs/BOGs/PTAs on the need to participate and sustain the BEIP. Ife's (2002) view on this is that any development intervention focused on empowerment must integrate educational mechanisms to raise the consciousness of disadvantaged people and enable them to meaningfully participate. Reiterating the need for, and the role of, sensitization in the BEIP, Mapya, an educationist said:

> I want to group stakeholders right away from the ministry, all the way to the parents on the ground: teachers … the beneficiaries and … the students. The involvement mainly has been: at the ministry level we organized for *sensitization* of the beneficiaries on the ground and mobilize[d] them on the need to support the project, so that they know the amount of money that is supposed to go to them, for what purpose ... how are they going to be ready to utilize these funds and also organize for sustainability ... at the end of it … Being a GOK/OPEC [partnership] there is need for sustainability of the project. (emphasis added)

Thus, the technocrats conceptualized consciousness-raising as a process through which information could be provided to disadvantaged people and to SMCs/BOGs/PTAs. The idea was to enable them to make informed choices about their participation in the BEIP. Nonetheless, consciousness-raising is more than just making a decision about whether or not to participate and how to spend aid resources. It is much more about the kind of participation and the extent to which the process increased ability of disadvantaged people to initiate political action on the structural and rights deficits they were facing. The way disadvantaged people participated in consciousness-raising processes is, thus, a critical measure of their level of awareness of rights and ability to enact these rights in the future.

As set out in the BEIP structure, political elites were meant to enact advocacy. Technocrats believed that political elites were more able to enact advocacy and affect consciousness-raising and participation of disadvantaged people because:

> These are people who are representing a wider range of population. That implies that the person ought to be an accessible person. You have access to them any time for information purposes. These are the kinds of people who are respected within the society. So … they can command some *authority*. We are talking about people who have some *influence* so that when you call for public participation, these people will be able to collect everyone in the community. We are talking about people like the political councillors, the members of parliament, the chiefs. These are people who are very influential. People who are commanding authority, people who are *respected*, people who have *character* such that when it comes to your interest of using them to reach to the community, because you are not known by everyone in the community, it is very easy to go through them … These are also people who can *communicate* clearly, may be with basic literacy level and who can communicate well because this is about passing information and if it is distorted, it messes the whole project. (Antoinette, emphasis added)

In tandem with alternative development (Friedmann, 1992; Ife, 2002), participation of political elites in advocacy was meant to guarantee 'political will' by linking disadvantaged people with the political structures that governed their lives. For this reason, when 'inviting' politicians to participate, technocrats chose those they considered actively involved in development of their constituents. The expectation was that the invited politicians would not 'politicize' the BEIP for personal gains.

However, contrary to technocrats' views that politicians ought to be impartial and apolitical, their practice of advocacy and participation indicates that schools and aid projects are arenas for political interests (personal and collective). On this view SMCs'/BOGs'/PTAs' testimonies reiterated that:

> This [interest] could be political … A politician in a given community when donor money like OPEC has come during his term, they expect next time to say this project has come during my time. So can you elect me? (Nashika)

Despite technocrats' awareness and concern to avoid practices that enhance personal as opposed to the collective good, the use of the BEIP as a campaign tool reinforced personal interests and redefined politicians as 'donors', not political agents. Such distortion of authentic political relations partly arose from the use of power-brokers to exert pressure on disadvantaged people to participate and enable technocrats to have secure access to, and acceptance by, these people. Such practices comprised integrity of the participatory process and depoliticized the political because rights (participation, access to information) by their very nature are political.

Acts of depoliticization were also revealed in the way technocrats invited politicians to participate and advocate for the BEIP in processes of participation and collaboration. As noted earlier, politicians used their authority 'as democratically elected leaders' to exert pressure on technocrats. To the extent that attempts to connect the 'personal and the political' (Ife, 2002) enhanced domination by politicians, it can be questioned to what extent advocacy 'conscientized' (Freire 1973) disadvantaged people about the structural disadvantages that inhibited their knowledge of rights and participation in development. How can we expect politicians to conscientize disadvantaged people about how the systems of governance from which they benefit perpetuate marginalities? This is not just a question about technicalities but also one of moral obligations. Benjamin, a head teacher, revealed that:

> Participation in this project has made our politician [active] ... particularly at school Y, this project has sort of brought some blessing politically. I am sure, after the political arena learnt that we had received some assistance from the OPEC people, blessings started flowing to the school. They also started allocating us money for something else such as renovating dormitories.

While some politicians participated by visiting the project sites, and spending their own resources to expand physical facilities in schools, this was not typically the experience. Others reported they had yet to meet their representatives. "[S]ince we started this project at our school, we have talked to the area councillor and she has never turned up even one day" (Maisha). As detailed in Chapter 6, structural disadvantages based on culture, gender and age mitigated against participation of some female leaders. The data are replete with views that some political leaders did not engage in the BEIP because they were busy in urban centres, while most project sites were in rural areas. This is to say regional disparities constrained the ability of political leaders to fulfil their advocacy and participatory responsibilities. According to an administrative chief, some political and civic leaders also failed to fulfil their obligations because:

> Sometimes when we want to participate ... [but] we are not informed about such things ... Sometimes those donors are not passing through our offices so by the time they come to the schools we are not aware. If we do not communicate [collaborate] with the headmasters ... we find it difficult ... to participate in the project. (Rosalind)

The way inadequate information amongst civil and political leaders and their weak social networks (or collaboration) with SMCs/BOGs/PTAs and technocrats influenced benefits of the BEIP mean that politicians and power-brokers cannot always be considered as direct channels for enacting consciousness-raising and participation of disadvantaged people. When enacting the structures on advocacy, technocrats believed that civic and political leaders were more able to use their political influence to affect participation and consciousness of communities within schools' neighbourhoods. However, as detailed later, their actions risked entrenching structural disadvantages either through exclusion or domination based on gender, culture and poverty.

Despite these shortcomings, the data reiterated that politicians should be encouraged to participate because, by so doing, they interacted with their constituents in developing schools. The experience gained from such interaction equipped disadvantaged people to demand responsibility from these political representatives when they privilege their personal interests over the collective good. However, the ability of disadvantaged people to initiate social and political action against entrenched structural disadvantage remained a distant dream because participation in the BEIP taught them that:

> Politicians should be there and should be encouraged to participate because they know more where these donors are and where funds come from than myself in the location. (Hamish)

This idea of establishing networks with politicians with a view to connect disadvantaged communities with donors misses the point. According to Ife (2002), consciousness-raising ought to enable these people to understand how structural disadvantages are perpetuated in society. The assumption is that due to the presence of oppressive structures and discourses, disadvantaged people have come to accept oppression as inevitable. Arguably, they remain unconscious of oppression, not just because they participate in it but also because the way they participate enhances technocrats' and donor 'complicity' in aid development (Kapoor, 2005).

The role of change agents is to raise the consciousness of disadvantaged people about the discourses and structures which inhibit them from effectively exploring disadvantage and oppression. As confirmed here, technocrats, SMCs/BOGs/PTAs and the politicians alike with the disadvantaged people require concientization. The data show that the way technocrats carried out advocacy represented disadvantaged people as passive recipients of aid resources, information and plans designed elsewhere, not as initiators of development. Such practice is also evident in the way technocrats implemented capacity-building and training seminars for members of SMCs/BOGs/PTAs and district implementation committees. According to GOK (2003a), technocrats used lectures with integrated questions-answer sessions and group discussions to build capacity within these groups and to sensitize them to the development objectives established through earlier processes. The 'knowledge' was tailored to specific project objectives in the areas of community mobilization, needs assessment, project management, school administration, financial management (budgeting, reporting, record keeping), accountability, procurement of stores and services, and development planning and integrity. It also emphasized the need for physical facilities to meet housing and health standards, and remain sensitive to the needs of the physically challenged. A training module was covered within five days through the "established communication channels" (Antoinette) of the education sector and the BEIP management structures, as shown in Chapter 3.

There was some evidence, on the part of technocrats, that the consciousness-raising in the BEIP was leading to an enhanced awareness of rights. Duncan confirmed this outcome by saying:

> When you hear some of these BOGs/PTAs … they are firing. People are not to be dictated to any more … initially we used to think elites are authorities of the village. He comes and dictates to the BOGs … [T]hese things are changing because people have become enlightened … they have grown out of this thing called 'nimewajua wale wamesoma' [I have connections with elites] … [T]hey [communities] may not be educated, but they are aware of some of their rights as citizens of this country. And that is a way forward for education ... [I]t is just about time change is coming … When you go to these district committee meetings, you will see [change]. People are talking more than they used to. Responsibilities have been pushed to the lowest level and that is opening their eyes. Aha! This is what we have been missing.

This concurs with Williams' (2004) view that a key determinant of emancipation is knowledge of human and political rights, and particularly the right of agency by the disadvantaged people. The data also show that participation contributed to processes of learning in tandem and increased awareness of needs.

> Participation has been to me an eye opener. It has built my capacity because in the processes of building others, I also learned from them … we in the ivory towers … think we know it all … wait until you go to talk to these locals and they tell you, No! Our needs are not what you are telling us. Our needs are … if you give us water, roads, electricity we will be fine. You go to a school thinking that you know their needs or what is best, only to realize you knew very little. To some of them classes may not be the issue … If you build a toilet for them even if you build a storey house for them, you are wasting time … It has been a learning process for me. We are talking about learning in tandem. (Duncan)

Capacity-building forums also provided space for SMCs/BOGs/PTAs to establish dialogical and collaborative relationships across class boundaries and share experiences of participation with technical experts from the Ministries of Health and Public Works. These ministries were also claimed to have increased their role in education as anticipated. These outcomes confirm Ife's (2002) view that participatory processes enable marginalized people to create new social networks and share experiences of oppression. According to participant observations, attendance in capacity-building was 100 per cent.

However, these outputs should not blind us from the exclusionary effects of the BEIP social-networks. To what extent did the methods of consciousness-raising and the social networks created affect dehumanizing structures and discourses that leave disadvantaged people powerless? When the approaches technocrats used and the said outcomes are juxtaposed with Brohman's (1996) concerns about who participates, how participation occurs, and where it occurs, it becomes apparent that in the BEIP, consciousness-raising least challenged dominant discourses and structural disadvantages. As said before, representation contributed to elite-to-elite networks in capacity-building forums. That is to say social networks between rich-poor and literate-illiterate at least in the short term were not achieved. This reproduction of existing social class divisions risked entrenching domination, because the very disadvantaged people who are most affected by structural disadvantages were excluded from capacity-building forums. According to SMCs/BOGs/PTAs, awareness-raising was a one-off activity in which technocrats were the mediums of knowledge, rather than a continual process in which disadvantaged people participate as creators of knowledge:

> The people [technocrats] who are used to introducing the project to the community, most of the times … do not understand the community … Somebody will be sent to sensitize you and he tells you he has two hours to spend tonight, and needs so and so. At the end of the day, the people who will [attend] are the people who are available then, … but not those who matter in decision-making in that community. And when they are confronted with questions, they end up giving the communities false expectations, they make false promises … The first question they are asked is 'what are you coming to do? What do you have for us?' So that they are accepted they will sweet-talk to the people. And then from there they will not be seen again. The ministry of education is notorious for sending different persons for the same project. Today you sent this, tomorrow another and the next day another person. (Jasmine)

The hurriedness with which technocrats conducted consciousness-raising in the perspectives of SMCs/BOGs/PTAs showed that "there is inadequate communication" and that they were interested in justifying their personal interests in aid because once they had "gotten their allowances … they don't consider the sustainability of that particular project" (Jasmine). Parsley, (technocrat) concurred with this view by saying that:

> We have had cases … where the money has … been spent in paying hotel bills for the same people day in and out to talk about project aims. In my view, benefits may not trickle down, or where we spent development partners' money [aid] to conduct community sensitization. The same people [trainers] … have recently been given a nickname, 'professional work-shoppers'. These are people who are in every workshop, be it is in agriculture, education, health and it all ends there.

Such practices concur with Mulenga's (1999) view that, to date, participatory development has contributed to manageristic systems, which are too costly and which have little, if any, impact upon the poor. As noted in the data, SMCs/BOGs/PTAs challenged technocrats' methods of capacity-building by demanding to be paid stipends just like the technocrats and to be given notes to take home rather than spend time in lectures to be taught what they could read by themselves, or what they already knew they could do better. Though underpinning a learning outcome on the part of technocrats and SMCs/BOGs/PTAs, such practices attest that consciousness-raising forums turned into 'sites' in which technocrats' top-down mindsets exerted bureaucratic pressure on the SMCs/BOGs/PTAs and disadvantaged people to participate. Yet, such domination appeared to enhance neither consciousness-raising, nor participation. Although all members attended needs-assessment, where attempts were made to sensitize them about their roles in decision-making and where they accepted to support the BEIP, their participation and that of neighbouring communities diminished during implementation. As said before, only the three (treasurers, secretaries, chairmen) out of a maximum of 13 members of SMCs/BOGs/PTAs who attended capacity-building forums from each school remained actively involved in decision-making. On this outcome Wamsha, an educationist, resented, "we [trained] them so that they could go back and sensitize the whole SMC/BOG/PTA so that at the end of the day each one of them is an expert, but not to go and be the only two or three people who are in charge of what is happening".

Paradoxically, this need to create experts testifies to 'new professionalism' (Chambers, 1994b). Professionalism meant that even where opportunities for participation were created, politics of representation, aid assistance and other structural disadvantages inhibited participation and the effective understanding of structural disadvantages by technocrats and disadvantaged people. These features imposed a technical approach to consciousness-raising which led technocrats to privilege their own knowledge over local knowledge and potential. Technocrats also rushed consciousness-raising activities because their focus on outcomes of physical facilities aimed to save teaching and learning time:

> We had said we were not going to talk [conscientize]. First ... we are going to the ground, build the classrooms and when we are about to be through, we shall start talking. I have considered that one as very practical because before we have done anything else, children are already in the classes benefiting. Although ... there was some minimum talk [initially]. We had to assemble the leaders and tell them what we want to do in the location [grassroots tier of local government]. (Parsley)

It is logical to build classrooms to ensure students do not lose learning time. These are important outcomes, given the BEIP's aim of increasing access to education by establishing an enabling environment for teaching and learning. However, this technical mindset compromised integrity of the participatory process by privileging outcomes of rights on the part of students over the democratic right of participation of their parents. This failure to balance between rights and obligations suggests that if parents are unaware of the structures and discourses which inhibit them from fulfilling their own rights and obligations to their children, it is unlikely the outcomes of awareness-raising would affect inequalities based on gender, culture, age and poverty (see Chapter 6). The next sub-section argues that this technical approach led to commodification of participation and devaluing of local potentials despite awareness that community-resources mobilization processes were critical to empowerment and transformation.

Community and Resource Mobilization

An approach to participation which technocrats used entailed mobilizing aid funds from OPEC and communities and their resources to support the BEIP objectives. A key aim of the BEIP was to build/rehabilitate classrooms and laboratories and provide sanitation and water facilities. According to official documents, these facilities lacked because:

> the government does not build schools or laboratories … Many parents and communities are not actively involved in management of schools due to lack of proper training in how schools are managed … Most learning institutions … lack supportive learning environments. Classroom conditions are poor. Furniture and equipment are inadequate or unsuitable. Water and sanitation facilities are unhealthy or non-existent … Many parents and communities are not sensitized enough on such issues and what role they should play in supporting education of their children. (GOK, 2002b, pp. 27–8)

To enable the government to fulfil its responsibility in planning and human development, technocrats secured a loan on behalf of disadvantaged school communities from the OPEC. The GOK aimed to use the aid funds to increase its role in building schools, a task which had largely remained in the hands of parents. Technocrats also enacted structures to enable SMCs/BOGs/PTAs to mobilize the disadvantaged communities they represented and use their knowledge, skills and social, economic and political capital to increase the impact of BEIP. Empirical data showed that:

> At the school level, we came to an agreement with the SMC … Since they will be the ones on the ground implementing the project, they would have to give bits of resources, but not in monetary terms. So they were to provide things like labour … Education administrators [were] to ensure that there is good coordination between the headquarters and the district level together with the committees who were involved in the implementation … There were so defined roles for each individual and I felt they [SMCs/BOGs/PTAs] were accepting their roles since we did not dictate the roles to them. We requested them to give us what [they wished] to do. So when they said they were … going to provide water for

> free, we felt that was participation enough. Others said they will collect sand … Out of all that, I felt the resources were very well shared in terms of the role for each individual. So it was not about just taking to them money because money is not the final end. It had to go deeper to the consumer getting involved in how to utilize the fund to make sure that they maximize the results of what was coming from the financial resource [aid funds] given to them. (Antoinette)

These arrangements contributed to different meanings of participation and development in the BEIP. As said earlier, by enacting structures for disadvantaged people to participate, some technocrats believed that they were "expanding democratic spaces of empowerment" in respect of these people (Duncan). This meaning implies a rights-based approach to participatory development, where the BEIP provided the context for disadvantaged people to actualize their right of participation and obligations towards educational development. Here, like Bushie below, most educationists described participation as a:

> … contribution or input by those affected by the project ... supporting the project [by] providing material resources such as sand, bricks, ballast, skilled, unskilled labour … [and] non-material support such as time [and] sharing ideas. (Bushie)

Likewise, SMCs/BOGs/PTAs saw "participation … [as] financial assistance from the government … donor and other well-wishers … parents working manually, like bringing water, building materials, giving guidance, monitoring and evaluation" (Jasmine).

These redefinitions of participation on a continuum from aid (technical) assistance to voluntary contribution emphasized a technical approach, which focused on providing solutions to the structural and rights deficiencies of the disadvantaged people. This emphasis on outcomes by maximizing benefits of aid valorized agency of disadvantaged communities, not as a right but as a method of enabling communities to save and redistribute surplus as in microeconomics and markets. Relieving disadvantaged people's burden of building schools conforms to beneficiary types of participation (Cornwall, 2003) where "the OPEC project has not come to incorporate our money to it ... the benefit comes as a supplement to our resources and abilities" (Reuben). Thus, by providing labour, materials and supervision of constructions parents reduced costs and could invest the 'savings' made out of the aid fund in other areas. These processes of 'saving' and 'redistributing' surplus attest to a micro-economic view that led to *commodification* of participation: participation was (re)branded and (re)packaged less as a right but more as an obligation through a discourse of 'using local potential' that in very subtle ways 'relieved' the government of its responsibilities whilst loading the same onto the citizenry. The commodification of participation is discussed further in the next section.

Commodification of Participation

Commodification refers, but is not limited, to processes (of decision-making, tendering, procuring/contracting, and competitive bidding) through which participation in the BEIP was weighed and translated into economic and human capital. It may also denote the actions through which such capital was sold (for example, employment/sale of knowledge/ideas/skills, labour, services), as is the case in market principles. Commodification draws from the view that the government (first system) has failed to achieve meaningful human development for the greater populace of the poorest Kenyans. The government can partner with donors and civil societies (second system) and citizens (third system) to double its synergies, overcome its weaknesses and provide sustainable development to all its citizenry. As research (Craig and Porter, 2003) has shown, this view represents a convergence of government, donors and citizenry politics within poverty reduction strategy papers and sector-wide approaches to planning policies from which the BEIP derived its legitimacy for people-centered, participatory and partnership approaches, in the first instance.

A classic example of commodification relates to the earlier said demands by SMCs/BOGs/PTAs to be paid a stipend when attending capacity-building forums. Contrary to technocrats' views of voluntary participation by providing 'free' labour, ideas and materials, commodification suggests that the BEIP provided opportunities for disadvantaged communities to gain employment and market their knowledge, skills and resources. An educationist recorded this view by reiterating that educating disadvantaged people on their roles was not motivation enough for them to participate. Rather, local resources, skills and knowledge should be valued because:

> When the contractor employs those local *fundis* [masons and artisans] they also benefit economically [through utilization of local labour force … skilled or unskilled … [and the use of] the materials used in the local area … (Benjamin).

These sensibilities concur with Ife's (2002) contention that, in addition to broad education to enable disadvantaged people to participate effectively, ownership and sustainable development are more likely where local potentials (knowledge, skills, resources) are used. The GOK satisfied this view by establishing structures that enabled SMCs/BOGs/PTAs to procure goods and services from the communities in which the BEIP was implemented. Valuing local resources was essential because "ownership … depend[ed] on how you have involved the locals" (Antoinette). SMCs'/BOGs'/PTAs' testimonies show that valuing local potentials enhanced self-esteem upon successful implementation of the BEIP, and "that is to encourage further development" (Benjamin). Using local resources to increase the impact of aid projects is essential to the extent that it allows 'spaces' for the disadvantaged people to participate and enhances their responsibility towards educational development.

However, where the aim is to reduce costs and spread benefits of the aid fund, using local potentials meant that under the guise of participation the GOK can perhaps reduce its expenditure on education, while loading the same on communities. On this view Sundukia said:

> One way of seeing participation is contribution of labour, it means they [government] will save in terms of money. If more people can participate and give their labour, they will reduce the cost and it will be better for them in terms of resources … participation to them means identifying needs. That makes their work easier too.

Such is commodification of participation because the GOK was able to reduce the burden on its part while transferring costs not just to the donor but more to the disadvantaged people. Indeed the disadvantaged people in the BEIP became triple payers. First, they built the schools on their own land. Second, they employed teachers where the government had not deployed these to the schools (GOK, 2003). Third, the BEIP was funded through a loan, not a grant. Here, participation is two-fold: direct tax, and in capital and kind through the project partnership. It can then be argued that through the discourse of 'valuing local resources' participation in aid projects enables the government to triple-shirk its own responsibilities while loading the same onto the citizenry. This is contrary to Ife's (2002) view of balancing between rights and obligations which, as said before, technocrats were not unaware of and which SMCs/BOGs/PTAs emphasized.

Rights and Obligations

Empirical data gathered as part of this study indicated that participants perceive that the government has an obligation to enforce responsibilities on the part of its citizenry, including the disadvantaged people. However, the methods of participation that technocrats used to enable disadvantaged people to meaningfully fulfil their rights and obligations in the BEIP and generally in life appeared to blame (or discipline) these communities for their conditions. Antoinette said:

> I [community, parent] am the beneficiary … It is not for the foreigner [donor] who is giving me money, not even for the … Ministry of Education. So every Kenyan should participate in these projects … In essence you realize that at the end of it all we will end up paying this loan all of us … not the Government. The Government will pay the loan as a facilitator … but me the local person, I am being charged the tax, which is being used to pay back the loan. In other words, whether I want to participate [in the BEIP] or not … at the end of it all, I will still pay back that loan through the little tax that I pay as I buy something in the shop. What I am saying is that you cannot exonerate [anyone from] participation

> … You cannot separate people from participating, because whether you want to participate willingly or not, at some point you will.

These methods underscore the point that aid projects must be legitimated not just by the disadvantaged people they affect but also by the greater populace of citizenry whose tax the government used to pay the BEIP loan. Thus, the government must enforce rights and create enabling environments for citizens to fulfil their obligations. The citizenry also have a responsibility to fulfil their obligations to the government (for example, payment of taxes by all including members of parliament, who are not passive but active participants in the development of their constituents). Reuben (a head teacher) emphasized the need for the government and citizens to fulfil their participation and development rights and obligations by stating that:

> If this project is in my community, I think I have a right to know what is happening in my community as far as the donor is concerned. So it is a right to know what is happening. Then I am obliged to … go to know [seek information] because you cannot know if you sit at home ... A right sometimes has to be fought for. So you go to fight for your right or to see what is happening. You are obliged to your right and you are obliged to promote it.

This awareness of the right of participation and other rights concurs with Ife's (2002) argument that rights and obligations are synonymous. To speak of obligations is to speak of rights. Again participation is a democratic process in which disadvantaged people struggle to fulfil their obligations and rights (Freire, 1972). The data show that disadvantaged people are willing to participate in development when they know their actions will make a difference (Ife, 2002). In this respect, Reuben (a head and science teacher) said:

> The ... community need actually is like a drive to make us participate more and surpass the challenges which we are undergoing. At the end of the day even if it takes more than ten years, the community need will still be there.

This understanding of needs/challenges as motivators shows that poverty and other disadvantages are not necessarily synonymous to powerlessness as the general participatory development literature suggests. The data show that the majority of parents willingly participated because BEIP helped to "enhance community development" (Rosalind) and people's wellbeing. Another head teacher also emphasized motivators and the socio-embeddedness of rights and obligations of parents, students and GOK, and the inseparability of rights and obligations of each (Ife, 2002) by saying:

> The factors that have encouraged me to participate ... I look at the end result of the project and the benefit [it] will give first to the student and the community at

> large. Even the country … If we are putting up laboratories and we want Kenya to be industrialized in the year 2020 … We [must] bring up young men and women who will at least contribute something to the nation of Kenya. (Benjamin)

The GOK's attempt to fulfil these roles contributed outcomes in the building of water tanks, classrooms and toilets in schools through the BEIP. These physical facilities were believed to have enhanced access to the rights of education, water, health, increased enrolment and gender parity by encouraging participation of the girl-child. The BEIP also enhanced enabling environments for parents and teachers to fulfil their rights of participation and obligations of expanding education. Considering the BEIP aim of enhancing teaching and learning conditions, these are critical outcomes.

However, given the earlier said donor and technocrats' interests in aid and markets and considering that community development, wellbeing and sustainable development entails control of resources and decisions (Ife, 2002), the viability of the methods used to enforce responsibility can be questioned both from technical and moral perspectives. Technocrats' focus on outcomes meant that participation is an obligation on the part of parents. By privileging obligations over rights, technocrats lost focus on balanced development, where the fulfilment of rights of students is dependent on the ability of parents to exercise their rights and obligations and the fulfilment of government responsibilities to all. For example, arising from the awareness of the interrelatedness of rights and obligations, technocrats enacted structures to increase participation of girls in education and women in the management of the BEIP. As detailed in Chapter 6, despite government efforts to increase gender parity in education and the number of women in school management, commodification of participation and representation, combined with cultural practices, encouraged decision-making through male representatives to deny women such opportunities.

This technical approach to participation risked entrenching inequalities by obscuring the rights and learning outcomes of the participatory process. As noted, despite the need to reduce cultural paternalism and increase gender parity, the technical approach used meant that culture and gender were only important inasmuch as they limited participation, not sustainable development. Arguably, technocrats' emphasis on obligations (technicalities, aid) over the moral (participation as a democratic right) dimension of development is denial of the right to self-determination and negation of both government and community responsibilities. Their focus on technical expertise and aid to address knowledge and resource gaps devalued local potentials and limited participation and alternative choices of SMCs/BOGs/PTAs and disadvantaged people.

For example, the focus on valuing local potentials would reasonably have been expected to encourage technocrats to draw upon local communities' conventional 'wisdom' regarding land and building/construction values. On the contrary, technocrats 'ignored' what SMCs/BOGs/PTAs considered the 'standard construction' that was acceptable to school communities in favour of the 'standard building plans' designed by architects and engineers (technical experts). Empirical data revealed that:

> There is the *standard construction*, which anybody else in the community is building. The *public works* have their *own standards* … to make sure that whatever building plan they [have designed] is used as per that standard of theirs. I think, *if we were given the freedom as stakeholders to do that* [design building plans] … *because we are building schools and structures everywhere and they are permanent* [we would have used our knowledge. But] … , saying it [building plan] has to be like this and that … until everything pitches on such standards [established by technical experts] … it is actually curtailing the involvement of the locals. (Mapatano, emphasis added)

Rather than increase efficiency and cost-effectiveness, in this case, technical expertise increased costs and accentuated vulnerability which Hamish, a head teacher, reiterated:

> These public works people [technical experts] … are really making things very complicated. They will always coat you exactly with the amount [aid] you have. There is not any allowance … something which you cannot even comment on … This is where our participation is limited … because this is conditional.

Technocrats' testimonies also contended that technical expertise devalued local knowledge by not allowing SMCs/BOGs/PTAs to use certain types of stone and other building materials within their proximities and which could have reduced costs. These testimonies confirm poststructuralists' (Spivak, 1985) criticism that needs of disadvantaged people continue to be observed and solved from the vantage point of external perspectives. Considering the worth of cooperative ways of problem-solving (Ife, 2002) and the need to avoid the dangers of localism (Mohan and Stokke, 2000), technical expertise has also a moral value, which was barely observed in the BEIP.

The data recorded that as a result of commodification of participation, technical experts made many trips to the project sites in large groups, used car hire rather than public transport, asked to be remunerated at higher rates than was agreeable with the GOK regulations, and preferred to be contracted. Such 'deals' enabled them to generate more monies than the reimbursements GOK offered. In these contexts, commodification of participation and the associated collaboration networks encouraged 'project money attitude' in the BEIP which, according to a technocrat, emerged because:

> in the past [aid] projects were rated differently … Different donors give a higher payment to those who are involved. But in this particular one, … it is the Ministry of Education which is going to pay back as a loan. Since we took it as a loan, we want all the money to benefit the child … Almost 100%. Operation money … is being catered for by the Ministry … in terms of going to train [or] monitor [the

> project]. So we will not really risk tak[ing] the [aid] money and pay[ing] people who are already on salary. (Carla)

The emergence of project money attitude just like the earlier said *mali ya uma* attitude as confirmed here derives from the need to increase individual, not collective, goods. Arguably, it is an immanent component of commodification of participation and the microeconomic view associated with it, where a core aim is to amass power to a few 'haves' (donor, technocrats and technical experts from the Ministries of Education, Water, Health, Public Works and Housing) from which to redistribute to the many 'have-nots' (disadvantaged communities (Arnstein, 1971). These ways of devaluing local potentials without providing permanent solutions to root causes of poverty meant that participation in aid programmes maintains balance in the status quo.

The next section builds on this view. It argues that though essential components of participatory development, monitoring and evaluation risked increasing control of the pace and process of development by technocrats and donors while entrenching corruption and dependency on aid.

Monitoring and Evaluation

According to the BEIP implementation manual, monitoring is:

> A regular, systematic and constant assessment of the progress achieved in the implementation of an activity … project. It seeks to establish the extent to which inputs, work schedules … targeted outputs and outcomes are proceeding according to the plans so that timely intervention measures are taken to correct deficiencies detected … evaluation … involves the process of measuring the performance of an activity … project to determine relevance, effectiveness, efficiency and impact according to set standards, targets and objectives. (GOK, 2003b, p. 59)

SMCs'/BOGs'/PTAs' testimonies indicated that to achieve these aims, the GOK sent technocrats and engineers fortnightly to the project sites to monitor progress. Notwithstanding the value of monitoring and evaluation, the practices of participation which such policies and activities generated appeared to devalue the organic development that Ife (2002) promotes. Indeed, joint monitoring activities represented technocrats' and donor attempts to control the 'process and pace of development' of the disadvantaged people contrary to principles of change from below. Such control manifested in the methods of participation. Sundukia said:

> The participatory methodology is like the process itself. Because you are building on people's responsibilities and … it is not like this idea of timing it, after two to three months I'll finish. It is a way of life … Once you have these

> values [belief in people's potentials and accountability] you make it a way of life that goes beyond the process … What is left with the people continues to be with them, whether a project is phased out … the people will continue to embrace those values. Unless people have those values, no matter how many PRAs/PLAs people do, it will just be like any other extractive method. If the people and facilitators have those values, then we can be assured of change and of course the whole issue of attitude.

The idea of building on people's responsibilities confirms (Burkey, 1993; Chambers, 1997; Ife, 2002) that participation is part of a development learning process whose knowledge gaps, methods and pace are determined by the communities. The attempt to implement the BEIP in phases and to disburse funds in instalments to test progress could be seen to negate control of the pace and processes of development by disadvantaged people while legitimating donor and technocrats' interests through monitoring and evaluation activities. SMCs/BOGs/PTAs said that these policies contributed to delays in the implementation of the BEIP. In other cases, the same policies enabled technocrats and technical experts to 'rush' the process of development rather than allowing it to 'evolve organically' (Ife, 2002).

Thus, monitoring and evaluation emerged as yet another 'manageristic system' (Mulenga, 1999) whose validity could be questioned to the extent that technical experts and technocrats were limited by their location in urban centres to be at the project sites when communities needed them. As detailed in Chapter 6, by not breaking these regional barriers participation risked reinforcing regional-to-environmental and class inequalities. It can also be questioned on the extent that disadvantaged people did not participate as expected. Some technocrats argued that some disadvantaged communities from North Eastern and Coast Provinces failed to participate in educational projects because they believed it is the responsibility of their elected governments to provide for those rights. On this view, technocrats appeared to use monitoring and evaluation to blame and enforce responsibility to these communities. Cautioning against blaming the disadvantaged where they do not participate as expected, Sundukia said:

> The whole idea of participation is based on premises of freedom and democracy … Human beings know what is good for them. They know at what times not … to engage. They have good reasons why they don't want to participate. I think the key question we should ask them is what are the factors for not engaging, or what are the factors that are promoting them to engage. I think that is the key area we should explore, but making the decision to engage, or not to engage I see that as exercising their own democratic right and it should be respected.

Thus, the challenge for the GOK to embrace participatory development practices in ways that empowered disadvantaged people to control their own futures still remained:

> Are ... [technocrats] willing to embrace the ideal of participatory methodologies? Even if you force people to embrace participatory methodologies and they just embrace it for the sake ... They say this is how we arrived at this ... it will be meaningless ... Anybody who is practicing participatory development and does not believe in the people themselves and their potentials, no matter how much you use those methodologies it is useless. Anybody who uses participatory methodologies and does not believe in transparency and accountability ... Even if a process goes through [offers outcomes] it is useless. A participatory methodology that does not believe in people's knowledge, that means the people have knowledge and can shape whatever you are doing ... believe that people can do it themselves; have mutual respect ... If you start seeing yourself as ... superior and they are inferior, that distorts the whole issue of participatory methodologies. (Sundukia)

Conclusion

This chapter aimed to explore how participation occurred in processes of planning, implementation and monitoring, evaluation and its impact on disadvantaged people. As part of achieving this aim, the chapter examined the objectives, methods, challenges/opportunities and some of the outcomes of participation. The official view included strong claims about increasing participation and the power of disadvantaged people to be able to control their own development. The data revealed a strong desire to allow the disadvantaged to drive the goals and direction of the BEIP through the needs assessment, consciousness-raising and community-to-resources mobilization strategies.

Although these aims and methods were enshrined in the official documents, in general, the practices of participation fell short of enabling disadvantaged people to achieve self-determination. This is because rather than treating participation as a process in which the disadvantaged people were constructors of knowledge and initiators of development, technocrats separated methods and outcomes (or rights and obligations). Such separation occurred through the use of representatives and technical experts who imposed a technical top-down approach to participation. This approach undermined integrity of the participatory process and principles of change from below, not least because the need to maximize aid benefits led technocrats to lose focus on the 'holistic' development they intended to achieve. It also led them to design methods which barely promoted participation as an empowerment and transformative process of development.

Consciousness-raising forums, for example, represented SMCs/BOGs/PTAs as passive recipients of resources and knowledge designed elsewhere, and excluded the disadvantaged people whose livelihoods the BEIP sought to construct. Commodification of participation in community-resource mobilization activities accentuated power and market shifts to the already powerful and greatly devalued local knowledge and potential. Though it remains an essential aspect of proper participatory development, under the BEIP monitoring and evaluation activities emerged as ways of controlling disadvantaged people's processes of development and entrenching corruption. Considering these outcomes, it cannot be overemphasized that promotion of technocrats' perspectives and interests remained significant inhibiting factors to disadvantaged people's participation and technocrats' effective understanding of poverty, culture, gender and other structural disadvantages. The next chapter explores these structural disadvantages under the themes of empowerment, sustainability and social change.

Chapter 6
Empowerment, Sustainability and Social Transformation

Introduction

The previous chapters have highlighted how the BEIP acted to exclude disadvantaged people from key decisions and planning processes. This chapter explores the extent to which the BEIP management structures, partnerships and participation impacted on empowerment, sustainability and social change. The central argument is that although the BEIP opened some spaces for participation and collaboration, the enactment and implementation of these features within an aid delivery system and through representatives and technical experts limited benefits of empowerment, sustainable development and social change.

This chapter reflects on the extent to which the BEIP affected social change. The first section considers the meanings of empowerment and impact on inequalities of class, culture, gender, age and poverty. The second section on *sustainability* examines the extent to which the BEIP challenged inequalities caused by broader environmental, socio-economic and political factors. The concluding section envisions possibilities for participatory development to offer 'total' emancipation.

As argued in this book, it is essential to understand the extent to which the BEIP structure delivered upon its promises. It is also essential to understand the extent to which the practices of partnerships and participation challenged dehumanizing structures and discourses of development which have to date remained dominant. According to Ife (2002), the three pillars of an empowering development approach are policy and planning, social and political action, and education and consciousness-raising. The structural and rights-based approach of the BEIP, as outlined earlier, engraved these ideals through the management structures, partnerships and participation. The aim of these features was to connect the political and the personal, and empower disadvantaged people to be able to make more accountable the political, social and economic structures that governed their lives. Ultimately, such empowerment and government responsibility was seen to enhance sustainable development and transformation. Although well intentioned, technocrats' perspectives and approaches to empowerment obscured effective understanding of structural disadvantages by themselves and the disadvantaged, and risked reinforcing inequality.

Empowerment

This section critically examines meanings and impacts of empowerment. The focus is upon the objectives, approaches, and the extent to which the resulting practices increased benefits of empowerment to disadvantaged people. Empirical data indicated that these communities participated in the BEIP because they wished to be empowered to control their own development and hold government institutions accountable. To this point, Reuben, in a focus group discussion, attested:

> We are participating because of empowerment: political, social and economic. Here, there are so many things we are fighting. Key on the agenda is the removal of ignorance. Upon ignorance, there is contribution to poverty, disease and … inability to exploit even the resources around, including the immediate environment. One cannot actively do that unless he/she passes through an institution of learning, where one is endowed with knowledge ... If the school does not have enough facilities that … will contribute to lack of knowledge, lack of information, lack of expertise in many areas … That tends to contribute to poverty. So when you have that institution [school] in place and students who are members of the local community, sent by the parents, they are bound to do very well within an enabling environment.

The data with technocrats and SMCs/BOGs/PTAs supported the belief that, by building institutions of learning, the BEIP enhanced enabling environments for the disadvantaged people to access the right of education. The official view was that, ultimately, education would empower disadvantaged people to overcome the deleterious effects of ignorance, poverty and environment and exploit their potential to enhance individual and collective wellbeing. While education was considered an important component of sustainable development, empowerment entailed transformative strategies in which the government and communities played their roles.

Hamish, a science teacher showed a desire for long-term rather than short-term solutions such as those realized in the BEIP by saying:

> We want to uplift educational standards … This [BEIP] is a benefit of education and that is why we are struggling all the more to put up institutions so that they can be accessible to the young. If the young do not have good facilities like this, they cannot change their way of thinking … I am not looking at this project endowed with [aid] resources. I am looking at that child, today and tomorrow and what the influence he/ she is going to make in that given environment, if she/he has accessed education and obtained good/acceptable standards … coming back to continue Community development … that is impact in that given community and that is where I work.

The point that community development is the ultimate goal for education concurs with Ife's (2002) view that the vision for empowerment is to increase the power of disadvantaged people over the political, social and economic institutions that govern their lives. It therefore meant creating an enabling environment to enhance the rights of access to education, health and economic activity (land, employment). Again, balancing education and other components of development is critical to the achievement of sustainable livelihoods. As such, empowerment also entailed social development as Reuben reiterated, "socially … it becomes easier to handle people who are well informed, well educated than when you are dealing with a lot of guys who are illiterate". These meanings of empowerment confirm the view (Ife, 2002) that communities are always engaged in their own development and that the role of government and change agents is to enhance enabling conditions for such development.

The extent to which the BEIP empowered the disadvantaged people is provided in the subsequent sections of the chapter. The next section shows that despite integrating an educational programme to reduce barriers that may have arisen as a result of cultural practices, *representation* cemented with cultural practices to exclude the very disadvantaged people who are most affected by cultural inequalities.

Culture

Generally, the data defined culture as people's way of life and as a set of norms. As said before, the BEIP aimed to impact on cultural practices that limited the educational opportunities of the disadvantaged groups and individuals living in city slums and ASALs. This definition and aim conforms with Ife's (2002) view of organic development where culture can either inhibit or promote empowerment and development. As a value system and as a source of knowledge culture ought to promote balanced development by ensuring access to rights by all. However, the data show that culture can be a significant inhibitor to the promotion of sustainable development:

> I would talk from the point of view of local communities because these are the ones who were really affected by socio-economic and cultural factors. Cultural factors … affect[ed] the project. They hindered it. There are some [communities] who do not believe in providing manual labour. They believe someone else must do it. That is a cultural factor that came out very clearly. There are some areas where we have nomadism. You even send the project but the user is not there. Culture is a limiting factor because the beliefs of the people influence … There are people who do not find a priority in that project … even after you do a lot of sensitization … Especially nomads who keep moving to other areas. They don't see why you are constructing a permanent building in their environment since they will move after sometime due to their lifestyle … There is also the aspect

> of attitude … of people looking at 'past projects have failed … what makes you think this one will succeed? It … took time … to convince them. That is a hindrance because [when] some … are saying 'this is our project', others are saying 'let us wait and see'. (Antoinette)

Disadvantaged people's low levels of awareness also help explain why participation in the BEIP was lower than expected. Cultural beliefs relating to manual labour, time and resources cut against the technocrats' expectations that disadvantaged people would contribute their own human and physical capital towards the BEIP. Indeed, the BEIP threatened some socio-economic and cultural practices upon which the livelihoods of disadvantaged people in ASALs depended. For example, one respondent noted that the BEIP challenged the dowry system:

> When this project was started, one of the strategies was to do community mobilization. I particularly did that. First of all the district sensitization was done by technocrats. Then I did divisional and zonal/village sensitization. Now, one pertinent cultural question that came up was that … , "you as a turkana girl, have you been paid for dowry? Our observation is, when girls go to school, dowry is not paid for them … That has impacts and far reaching consequences both for my family and the community … In my community if you have been paid for dowry, a woman is considered … a proper woman … The dowry … benefits members of the clan … If I have not been paid for dowry, it is like … You make them lose the opportunity … You make them lose an economic base. Then, I am not recognized in that community. It is like going to school because of that [BEIP] initiative, you have been detached from your community/clan. So there are two types of people created here. A turkana girl who is foreign in quotes has gone to school and then the turkana girl who remained in the traditional lifestyle … The thing is, these people felt and … still feel us who have gone to school are not proper role models. They even say … , 'when our children ... go to school, they settle in towns and leave us without care … You just hear they are married by other people they have never seen. We are not being told [or consulted]. So why should we take these children to school ... It is like the family institution, the clan, the traditional community is threatened. (Jasmine)

The view amongst some disadvantaged communities was that the BEIP (and the education system in general) was a threat to the socio-economic institutions that held families and communities together. That the BEIP did not offer acceptable alternatives limited the efficacy of the intervention to achieve empowerment. Moreover, the approaches technocrats used to achieve the aims of empowerment were perceived to further marginalize rather than emancipate the disadvantaged people:

> I want to give an example of the GOK/OPEC project ... The people who are in the planning and who have been sensitized are … SMCs and … project

> coordinator[s] … There is also a district team that was sensitized and trained … The thing is, this SMC is only a representation, a small portion of the community, or that particular village where the school is built … The extent that these SMCs … may not be people who are popular who can convince the community to give their contribution or give labour that kind of thing … that is a limitation … A weakness is the little representation … In the four [primary] and one secondary school they are assisting in the district, there is one … which has lagged behind. Some schools have reached windowsill others have roofed. That school [it's now they are beginning] ... making bricks … You wonder … It is like this community was not sensitized. (Jasmine)

Despite integrating a consciousness-raising programme, the enactment and implementation of the BEIP through representatives (SMCs/BOGs/PTAs) and technical experts meant that disadvantaged people themselves did not participate as expected. Again, although SMCs/BOGs/PTAs attended, they were not seen as creators of change, but rather as recipients of development (aid assistance). Such perceptions inhibited empowerment and social change benefits that disadvantaged people could have had from these educational processes had they actively participated. Furthermore, the project was driven by the aid agencies. The content offered in capacity-building seminars was tailored towards project management and how to mobilize communities to participate, as a way of maximizing the benefits of aid rather than creating awareness about how cultural disadvantage limited access to rights of education and participation. However important, this education fell short of reaching the standards that, according to Ife (2002), are needed to address issues of how cultural and other structural disadvantages inhibit the ability of disadvantaged communities to claim the spaces created for them to participate. SMCs/BOGs/PTAs suggested that such broad education was more able to address cultural practices that have, to date, denied women rights to own property, which promote forced marriages and female genital mutilation (FGM) for girls, and which make parents deny girls access to education.

Despite these shortcomings, like other aid development interventions the BEIP was creating awareness and impacting on socio-cultural change:

> The 'doctors' of FGM are being told that even in child[ren's]-act it is written that if you do such kind of practices and something bad happens to the child, you can be imprisoned. So culture is being affected by religion and know-how of the people. As people grow close to religion, they know which culture to take as indicated in the books that they are referring to … Bible, Quran, and Children's Act or rights conventions. I have in mind the World Vision, Moyale Branch that is creating more awareness of cultures that are not suitable. (Shirikiano)

This redefinition of cultures through education, religion, human rights conventions and children's acts underlines an important convergence of aid development through government partnerships with civil society, donors and local communities. According to Ife (2002), any development intervention underpinned by an empowerment system must value local cultures inasmuch as these adhere to conventional rights. To maximize their impact, change agents must thus connect the personal (spiritual) and the political with a view to empowering the disadvantaged to overcome the deleterious effects of localism (Mohan and Stokke, 2000) and the imposition of dominant narratives where human rights are stated in 'universal' rather than contextual terms (Ife, 2002).

As noted, in the BEIP, technocrats' approaches either devalued local cultures or privileged the individual rights of children over the collective rights of communities. This tendency led SMCs/BOGs/PTAs to argue that compared to the GOK service delivery systems, civil societies are rather more able to allow disadvantaged communities to originate strategies to address their needs without interfering with cultural systems:

> … in my district, NGOs, development partners [government and mult(b)ilateral donors] have formed a system whereby they go down to the community and form CBO systems. Through these CBO systems they are able to address issues of development without necessarily interfering with the cultural systems of the community. In the CBO system … people are given a challenge to come up with their own programmes. Culture and religion are distinctly different. Some communities may use culturally based programmes, others use religion. It is easy for someone to change me religiously and I may not change culturally … So CBOs have assumed that system of religion … that is what I have observed. (Shakombo)

These convergences between religion and culture, and government, donors and civil society are critical to our understanding of how the BEIP partnerships affected the local cultures of the disadvantaged people. In an attempt to strengthen partnerships, the GOK established a management system where the donor, technocrats, political elites and civil society (SMCs/BOGs/PTAs) were allocated roles to play in the BEIP. A core requirement was that SMCs/BOGs/PTAs must have a sponsor. In most cases, this was a religious organization. The data point to an awareness of the 'fluid' nature of culture. Since culture is dynamic, a partnership between the government, civil society and the disadvantaged people themselves was able to address the developmental and rights challenges of the BEIP. In part, one problem with partnerships based on religion and culture is that people may embrace new religions or cultures without necessarily changing the practices which inhibit access and promotion of rights in the first instance. The efficacy of aid projects enacted through government partnerships with donors and CBOs/NGOs to impact on cultures that inhibit access to rights can thus be questioned. These organizations are taken to represent the interests of disadvantaged people. Yet, representation in the BEIP led to exclusion of the disadvantaged people who are most affected by cultural inequalities. For this reason, participation through representatives cannot be considered to directly impact on cultural challenges faced by disadvantaged people because:

> … in my area we see culture and religion as synonymous. The dictates of religion are more powerful than cultural dictates. We have reached a stage where we are not able to identify the difference between religion and culture … from the works of the GOK and development partners … The issue of CBOs is created within us, but it has not yet matured … They [CBOs] have been formed but they are not operational … These organisations are desperate in accessing funds. They are not better than the communities they are trying to assist. They have no donor and they have not been formed with the consultation of the community in total [or by] … its [representatives]. Now it is not addressing the problems of the community. What is to be done is the issue … My community in particular, what they need is awareness. (Pakomosa)

The approaches technocrats used to address the deleterious effects of disadvantaged communities' cultures were a significant hindrance to emancipation and effective understanding of structural disadvantages. The failure by technocrats to legitimate management structures, either through consultation or election processes, meant that disadvantaged people did not consider the BEIP as representing their interests. For example, despite the BEIP having a strong focus on rights, the creation of awareness through representatives and technical experts who did not identify with local cultures and religions led SMCs/BOGs/PTAs to label these facilitators 'professional work-shoppers'. These labels highlight some critical questions regarding the efficacy of the methods of empowerment used in the BEIP. Where local facilitators were used, for example among the Gusii in Nyanza Province, they were not considered role models because either they had done FGM on their girls or were seen to be motivated by personal interests in aid and markets. For this reason a parent asserted:

> … we [community] have not come to the reality of who we are ... we do not have women who are ready to represent the government fully from their heart, ready to change the system of FGM. So these are the people who have made the system to continue for so long … [when] they go to the workshop and ask other women to stop, they are asked 'have you done it to your children?' If yes, who do you expect to change? You should have served as a role model so we could see the goodness in it … The NGO is better because they [use] people who have not been mutilated … These are better role models. But if they (GOK) continue to use women who … have been mutilated and have mutilated their girls … who are interested in money … It is not easy for the community to change their attitude because they know who you are ... That is the problem. (Omambia)

This view that change agents both in mainstream government and civil society were motivated by their own interests in markets and aid, rather than cultural needs, confirms Spivak's (1985) contention that disadvantaged people's development continues to be observed from external viewpoints. A core aim of the BEIP was to build classrooms and laboratories and provide water and sanitation facilities. As shown in Chapter 5, these are indeed important outcomes. However, some communities in Coast Province could not use toilets constructed by foreigners because that was culturally inappropriate. They would rather construct them themselves. Establishing permanent classrooms also risked forcing people to change from nomadism to more settled lifestyles. Such practice problematized people's lifestyles and imposed undesirable changes. One district committee member said:

> I believe we should address issues and create awareness. In my community the issue of water is their problem and you find everybody talking about it. Indeed they tell you, you are talking about education, you are putting so many efforts you are building so many classrooms, yet the lifestyle is that which is highly mobile. They have a nomadic lifestyle, but the facilities you put in place are static. You expect somebody who has gone 50 km away to access that facility. What we are saying is, we should have a system that takes services to the people but not the people to the services. This is what is happening. We want the people to come to the services, but the services should be taken to the people. This is the way I see it. (Oromosa)

Technocrats' testimonies concurred with documents (GOK, 2005a, 2005b) that the government was implementing mobile schools in nomadic communities and night teaching-to-learning sessions in ASALs in an attempt to take services closer to these people. Nonetheless, the data are replete with testimonies that a broader system of education and consciousness-raising than was provided through the BEIP was needed to emancipate the disadvantaged from cultural inequalities. A broad civic and citizenship education that focused on enhancing awareness of human and political rights was envisaged to be more able to reduce cultural paternalism.

The next section on gender shows that without such education, the efficacy of participatory development to affect cultures that treat women, children and youth as inferior and not able to influence community development was limited.

Gender

This section argues that representation (or technical expertise) risked cementing cultural practices that exclude women from decision-making and leadership roles even where affirmative action policies are put in place. According to Ife (2002), affirmative action policies are essential to increase access to services and participation of disadvantaged groups such as women, the aged, youth, children, physically and mentally challenged, and the poor. The training module that technocrats used for capacity-building (GOK, 2003a) and the project implementation manual (GOK,

2003b) stated that all treasurers who assumed management roles in the BEIP must be women. This affirmative action conforms to a broader policy where one third of SMCs'/BOGs'/PTAs' membership positions are reserved for women (GOK, 2005a, 2005b). It aimed to facilitate participation of women in management of the BEIP and also affect gender inequalities in the SMCs/BOGs/PTAs, which were considered to be dominated by men.

In practice, however, this policy floundered delivering the desired outcomes. The domination of SMCs/BOGs/PTAs by men and their attitudes discouraged women from participating in the BEIP. In support of this point, Hamish said:

> Gender goes hand in hand with attitude ... For example, if we are to look at the number of women who come and participate in implementation of the OPEC project, if those women look at the composition of the committee that deliberates on the implementation of that project, they find that they are all men. There are some women chauvinists who might feel that they [men] have put us aside as women, so why should we go and [participate] when they never even considered us when they were selecting the members.

The cultural belief that women are less intelligent than men also limited the effectiveness of affirmative action policies aimed at addressing gender inequalities. According to one teacher, such inequalities "can only be overcome in due course, when women are involved in participating in issues of decision-making and with them contributing to ideas. If someone is giving out an idea, you go for the idea regardless of whether one is male or female" (Rodham).

The perspectives of gender held by the technocrats, SMCs/BOGs /PTAs and the disadvantaged people themselves limited participation of women. Women were not able to take up their managerial responsibilities, because in most disadvantaged communities there were few literate women to take up such roles. It follows then that, by enacting the BEIP through representatives who had to satisfy minimum academic achievement levels, technocrats risked reinforcing inequalities of gender. The data recorded significant relationships among cultures, low literacy levels and the participation of women. One science teacher said:

> Can we correlate that [women representation] with the level of illiteracy in a given community? Illiteracy that goes with tradition and culture ... If you talk of community involvement in the OPEC project, there comes in that aspect of culture. This should be done by this particular sex. This is work for women and this is work for men. So community involvement there becomes a problem. (Chweya)

This correlation between gender and sexism and culture and low levels of awareness led SMCs/BOGs/PTAs to be concerned about how the language they used represented women and the disadvantaged. Reuben, a teacher, said, " ... maybe we may not call it illiteracy. We can call it limited literacy". The concern for discourse, sexism and disadvantage arose from the view that certain beliefs, traditions and cultures skewed participation towards a particular gender. Where communities believed that a type of work was meant for men, then when called to participate it was more likely that men participated in the ways that were considered culturally appropriate to them.

Where tasks were culturally associated with women, it was more likely that when called to participate, only women took up such tasks. For example, SMCs/BOGs/PTAs were supposed to appoint women to be treasurers of the BEIP funds and elect them in the procurement/tendering committees. A key aim was to increase the role of women in decision-making. These opportunities notwithstanding, most women in ASALs did not take up these managerial roles as anticipated. This failure to participate led some men to argue that women "do not take their work seriously … " (Hamish). In response, a female participant in the focus group asked:

> Is it women who are not serious or is it men who do not give women the chances for them to participate? If women were given the chances, we would have more women in this activity. (Rosalind)

Participant observations showed that the meeting was attended by one female. It was also observed that some male participants were somewhat 'violent' and uneasy when Rosalind spoke as shown in the response "If I give you and you do not even come to the tendering committee, how do I even give you more chances? We are wasting time" (Hamish). As noted, house chores may have limited women from taking up management roles. Nonetheless, there is consensus in the data that men denied women such opportunities because culturally, decision-making is not women's responsibility.

Further evidence showed that men were not unaware of the deleterious effects of cultures that vested decision-making authority on men as Dororomo, an educationist, said:

> In the communities and that is according to the customs, roles are very specific and are given to both genders right from childhood to adulthood and even old age. As much as … there is no conflict between the roles of the two genders … The most unfortunate thing is that there is no fairness … The roles given to women are much more compared to those of [their male] counterparts … The men are the ones who run how/what each gender does … The men have good time in the sense that they do very little compared to the ladies. But the ladies seem to accept because they do not take part in decision-making … [which] is completely the role of men.

These testimonies acknowledged that the BEIP provided opportunities for women to participate in educational management. It recorded optimism that participation in SMCs/BOGs/PTAs even after the BEIP was completed would progressively enhance benefits of empowerment and social change for both men and women.

However, the view advanced by some men that women expected to be denied leadership and decision-making roles suggests that the men resisted policy changes that offered women such opportunities. Dororomo reiterated:

> Despite that rigidity of men not wanting to have complete change, things are moving for the better. But still we have a long way to go because children's preference for boys is still there … Even the mother prefers the boy … In terms of [leadership] roles, education, inheritance of property, the boy child is given the upper hand. So we have a long way to go to ensure that there is a level ground.

Due to such entrenched cultural paternalism women have arguably become 'unconscious' (Ife, 2002) of cultural and gender inequalities and have come to accept a denial of rights to own property and have leadership and decision-making authority as the norm. For this reason, women also seemed to prefer male leadership. This behaviour led men to contend that women are 'enemies' to themselves. However, such blaming was negated by the view that even men preferred male leaders and hardly suggested women for civil elections. While women appeared to have accepted their lower status positions in communities, the data are clear that: entrenched paternalism and cultural practices that exclude women from decision-making in preference to men encourage the belief that women's voices should be heard through male representatives; promote the notion that women are less knowledgeable compared to men; enhance the cultural view that women are good as housewives – their role is in the kitchen; and suggest that women are the weaker sex and thereby require men 'shields' all of which have led women to devalue their own 'humanity' (they feel inferior) and potential. The paradox is that the men gave women opportunities to participate in SMC/BOG/PTA activities claiming that women had the time and courage to work for the community. This is a compliment and an opportunity for women to participate. However, the men retained decision-making authority in much the same way as the national taskforce did.

Thus, cultural practices of decision-making by men reinforced by the technocrats' decision to retain decision-making authority with the national taskforce combined to exclude women from decisions and management of the BEIP. This attempt by technocrats to prescribe gender representation and retain decisions to satisfy dominant structural and cultural practices risked entrenching gender inequalities.

While affirmative action policies are important to facilitate the participation of women and a strength of the BEIP, the requirement on literacy levels made it possible for SMCs/BOGs/PTAs (which were dominated by men) to invoke cultural beliefs and the low literacy levels to preclude women from participating in management. The literacy requirements negated the benefits of empowerment and social change in communities that preferred to educate boys over girls because women either did not have equitable literacy levels with men, or the minimum levels of education prescribed. Such conditions, together with the rigidity of men, limited empowerment benefits and risked cementing cultural and gender inequalities.

Moreover, while women took up management positions in the BEIP, their contribution to decision making remained low because men chose "not to listen" to women's ideas in SMC meetings (Hamish). As such, a key determinant to the empowerment of women is to address the cultural barriers which led them to be treated as lesser beings and which, as a consequence, make them feel insecure. The data showed gender relationships which are also far more complex, but for a book of this size, it is necessary to make only general comments. Increasing access to education and improving sanitation in schools was believed to increase the security and comfort of women and girls. However, to reduce gender inequalities both in schools and educational development, men need to unlearn the cultural beliefs and practices that make them devalue education for girls and not listen to women's ideas. One belief to unlearn is the patriarchal attitude that women should aggressively *fight and compete* with men as equals. Next, is the tendency to invoke culture to deny women/girls decision-making and education rights. Third, is the interpretation of gender 'purely' in either masculine or feminine terms.

The convergence of these cultural perspectives with pluralist and elitist ideologies in the BEIP meant that where women are unable to compete favourably either male 'shields' or elite representatives were needed to put through women's issues. The paradox is that these perspectives led men to perceive the empowerment of women as a threat to their own superiority. The use of words such as 'superiority', 'aggressive', 'let them fight and prove that they can deliver' in the data showed that some men perceived attempts to address inequalities of gender in terms of increasing domination by women. While acknowledging the role broad education played in reducing gender inequalities and enabling women to progressively increase their role in leadership, as is the case with pluralist and elitist ideals (Ife, 2002), men still expected women to compete with them for leadership and educational opportunities as equals.

This convergence of representation and competition with cultural and gender perspectives meant that men could easily justify uncooperative actions against women while blaming them on culture or biological differences. Where women were expected to compete, rather than effect cooperation and gender inequalities, affirmative actions were more likely to reinforce the status quo. Marijuana and alcohol taking by the men was perceived to negate their responsibility and precipitate violence against women, children and girls. Violence manifested more dramatically in communities that traditionally encouraged decision-making by men, especially among the Maasai of Rift Valley Province and Gusii of Nyanza Province. For example, among the Gusii, most schools were believed to be dilapidated because male domination inhibited effective participation in SMCs/BOGs/PTAs. Despite being more entrepreneurial, the cultural exclusion of women from property and land-ownership decisions meant that men did not expect them to undertake decisions relating to construction and other property related issues in the

BEIP. The problem is that the way the BEIP was promulgated appeared to cement, not to challenge, such violent behaviours and gender inequalities. Representation and competition meant that gender participation was about integrating the few elite women who were able to compete favourably in a system that was dominated by men. These features excluded the majority of less literate women and isolated the men who were unable to compete favourably.

The competitive behaviours that emerged from the practices of participation and partnerships were a good recipe for violent, as opposed to peaceful and non-violent ways of problem-solving, which Ife (2002) recommends. As noted, the consciousness-raising programmes empowered women with skills and attitudes needed to effectively participate in SMCs/BOGs/PTAs. However, to the extent that the men continued to make decisions in private arenas, the quality of their participation remained elusive.

The data are clear that women representatives presented in SMC/BOG/PTA meetings the decisions and ideas to which their husbands consented. At face value, this could be justified on grounds of cooperation. Nevertheless, these practices were unlikely to challenge male domination and gender imbalance because:

> ... culture, tradition, that kind of background ... seems to be following them [women]. Even those who are educated, even when they are put in positions that they should influence, they still seem to go back to [culture] instead of letting go, because by taking that position, men are happy with them. They see them like real women while those who go against, or those who really assert themselves and bring out the issues that affect women and or be a real voice, they are seen as radicals. Many women don't seem to like to be seen as radicals because that tends to exclude them. Somehow the voice of women is not as it should be. If you look at the programme of education, women are worse affected. Yet their voice is not as much [laudable] as the problem. There is need for a lot of affirmative action, a lot of unlearning for both men and women. (Sundukia)

The chapter will return to the point made in the data about the need to 'unlearn' traditional practices. For the moment, it is worth noting that these practices attest that both men and women are unconscious of the factors that limit them from effectively understanding gender inequalities and exploring possibilities for social and political action. Where such understanding is demonstrated through the creation of affirmative action policies, perceptions about representation, technical expertise and bureaucracy obscured benefits of empowerment and inhibited gender parity.

Contrary to Ife's (2002) view that affirmative action policies should challenge dominant perspectives by assuming cooperative as opposed to competitive methods of problem-solving, competition and gender imbalance were not just outcomes of the BEIP but were entrenched in the education and media systems. Cultural paternalism and domination were entrenched in families, schools and at national-macro development levels. The issue here is that technocrats used approaches of representation and competition that were perceived to entrench the very cultural and historical practices of domination and exclusion that were inherent within the broader education and political systems and among SMCs/BOGs/PTAs.

To affect such gender imbalance and cultural inequalities required a broader education system than that provided in the BEIP. Where women were said to suffer from 'cultural syndrome', cooperation – not competition – is more able to empower them and increase their roles in decision-making. The next section builds on this point by arguing that where decision-making is premised on perspectives of experience and 'universal' wisdom as promulgated in the BEIP, representation risked reinforcing cultural practices of exclusion and denial of rights on the basis of age.

Age and Experience

According to Ife (2002), age is a significant defining factor of structural disadvantages, participation and empowerment. The data showed that participants categorized age roughly into groups of adults, youth and children. Moreover, there was a very fine line between these age brackets owing to the socio-embeddeness of rights and responsibilities. Age significantly affected the relationships of partnerships and participation. These relationships drew upon the way the bill of rights (constitution), local cultures, and professional milieu defined age.

The one which derived from the constitution considered a child to be below 18 years of age. Above age 18 people were either seen as youth or adults with citizenship rights. An important point to note is that children do not have direct access to citizenship rights, but have access to such rights through parents. The cultural defining factor of age varied according to communities. Most communities grouped members into age-sets (or sex-sets) comprising children, youth (those who have undergone rites of initiation and are ready to marry), young adults (newly recruited to adulthood through marriage), middle aged adults (in transition/being prepared for leadership roles in communities), and adults (from among whom political, social and economic and religious leaders were selected). The work related factors defined relationships (e.g., student–teacher, employer–employee, government–citizen and parent–child).

These perspectives are critical to our understanding of how technocrats promulgated participation and partnerships and the way the BEIP impacted on structural disadvantages arising from age. When SMCs/BOGs/PTAs were asked to relate their experiences about how age influenced participation and partnerships in the BEIP, a teacher in the Coast Province responded by saying:

> I think when you are old you should be left out because even when we come to meetings like this, they are slow. (Chamkwezi)

The view is that 'old' people process tasks and information slowly and were unable to participate actively in the management of the BEIP because of low attention and memory spans. Old people were also perceived to be physically too weak to provide manual labour in the construction sites as prescribed by technocrats. To the extent that these disadvantages were potential reasons to deny old people their right of participation, the BEIP risked accentuating inequalities caused by age.

SMCs/BOGs/PTAs were aware of the dangers of blaming old people for their condition and excluding them from participating. A teacher, in reference to the relationship between age and experience, said "Amekula chumvi nyingi tumsikize" [a Kiswahili proverb meaning, what old people say/do should be adhered to because they have loads of experience] (Kasim). Apparently, this idea of equating age with experience (or wisdom) is not just an element of local cultures.

It also influenced the formation of the BEIP management structures. As said before, technocrats believed that their most senior colleagues were more experienced and able to enact and implement feasible relationships and policies to empower disadvantaged people. This equation of age with wisdom and experience with power led to a conflict of interest and inhibited development. Hamish said:

> That is why we have found in some communities they [old people] will water down what has been spoken by us, young men … 'what will this young man tell us?' I think age has also something to contribute towards participation in educational projects … That is what has hindered a lot of development. Elderly people think that they are everything … When we young people say we want to do this, they say, what do you know you young people? We have been here for ages and we know what you need [experience]. Until sasa [now] we say we [young people] have to go by force ... I remember an experience we had in our primary school … [Participants] are going to bear me witness. There are some elderly people who did not want to come out of the SMC. They wanted to remain there and do things the way they did last year and ten years ago … So we said no and now we changed. The youth came in and good luck the youth took over the leadership of the school. At least in one year things in the school are changing … There should be a [better] way of dealing with the old generation to have changes.

To the extent that age and experience inhibited change, we need to question what knowledge claims are more able to enhance agency and, in the next section, a sustainable livelihood of disadvantaged people. The point that aged people devalued the knowledge of young people and denied them leadership positions based on claims that they are inexperienced has implications on the way technocrats enacted the management structure of the BEIP and the impacts it was perceived to have on the disadvantaged people. Experience and age were the main reasons why the national taskforce retained authority to make decisions. The formation of the taskforce drew upon the view that members were most experienced in their areas of specialization.

This attempt to use age and experience to legitimate the status quo limited the participation and decision-making rights of SMCs/BOGs/PTAs and the disadvantaged people. While experience is an important element of development management, the assumptions that the most senior bureaucrats made the most effective decisions and policies and that the time for the youth to lead and participate in development was in the distant future contradicted the BEIP aim of enacting partnerships and participation across age, cultures, and the social, economic and political spheres of society.

Although partly responsible for legitimating bureaucracy, age and experience were not hindrances to participation and partnerships in themselves. The use of age and experience to deny youth leadership roles and retain control without necessarily satisfying empowerment and social change interests of the disadvantaged people and the ones considered 'young' was the problem. The data underscored this unwillingness by those with power and in leadership positions at the school, district and national levels to leave office and train others for management succession as a key hindrance to the meaningful participation of middle level managers, students/children and the disadvantaged people themselves. Contrary to Ife's (2002) views on peace and non-violence, this inability to cede power and lack of cooperation by the aged leaders contributed to competition. It also led the young to resort to non-peaceful strikes, eviction and demonstrations to claim what they perceived as a denial of their rights of participation and decision-making. The data indicated awareness by technocrats and SMCs/BOGs/PTAs of the deleterious effects of these features and the desire for cooperation and a peaceful means of problem-solving. Reuben, a head teacher, suggested:

> We should have a few [old people]. We should not leave them [out] completely … [Let old people play an] advisory role … Some of us [are] old but we are young in that we would like our children to live good livelihoods … Old people should come aboard but … they should not assume that they are capable to do everything.

This view that the youth should not claim their rights of participation and leadership positions by completely excluding the old concurs with Gandhi's (1964) challenge that change agents should promote the change they wish to see in the world. It also conforms to Ife's (2002) contention that unjust means cannot be used to achieve justice or wrong means used for right ends. Implicitly then the understanding of age and experience in terms of the 'vision' leaders hold for the wellbeing of the present and future generations is a strong basis for forging cooperation across different age groups and for cultivating effective understanding of the deleterious effects of age.

Technocrats and political elites were believed to be the 'vision-bearers' within the BEIP. It is for this reason that technocrats created representational roles for them in the management structures. The paradox is that where SMCs/BOGs/PTAs and the disadvantaged communities appeared to question the logic used to enact participation and partnerships, technocrats treated these behaviours either as animosity or resistance. The data testified that ownership, sustainability, and empowerment depended on how vision-bearers promoted participation. Where communities are sensitized and involved in the project at the design and planning levels and, where continual cooperation and dialogue prevailed, communities rarely resisted.

Despite intentions to promote ownership and empowerment, the attempt to retain decision-making authority with technocrats and to use their knowledge as the unquestionable norm negated benefits of ownership. Again, the use of political elites to enact advocacy, influence the participation of disadvantaged people and infuse political will in the BEIP processes appeared to reify age and experience even where these groups were seen to significantly lack in vision and promulgate their own interests rather than those of the disadvantaged people. Antoinette avowed:

> Multiparty government environment is a hindrance. Others are not visionary … They lack the vision. They are not supportive to that [project] …

Where age and experience served the interests of the status quo, the BEIP risked cementing inequalities of culture, class, bureaucracy and as detailed under sustainability, denial of rights due to political marginalization. SMCs/BOGs/PTAs argued that age and experience made those who are already in leadership positions think that they have a monopoly of knowledge. Yet "the young have knowledge. [The aged] should not keep on talking of what they did thirty years ago … life has changed, everything has changed. [They should give the youth a chance] … to headship positions" (Mapatano). That is to say, age and experience are not reason enough to cling to power and leadership positions. The issue of age is highly contested because both culturally and bureaucratically, age is seen as a sign of wisdom, not just experience.

To assume that young people should hold leadership positions, as heads of schools and departments with or without experience, is to challenge the very wisdom that underwrites certain cultural, departmental and organizational knowledge claims in the education sector and in all government and private institutions. By reifying experience and age, technocrats undermined the BEIP's aim of increasing access to the rights of participation and decision-making on the part of disadvantaged people and of students. When SMCs/BOGs/PTAs were asked how students participated. Benjamin, a head teacher, said:

> To say the truth … our students have not been involved. We are putting up a laboratory. They see a construction going up. That is why at the end of the day, they go writing [tagging] and they burn it ... they don't call it their own development. In every writing they … always damn instructions … We put up a primary school next to secondary school Y … [We] spent 4 million shillings worth of blocks. We never called the children to tell them, the donor came and brought the money and we put up a beautiful block for you. They are sitting

> there learning, but it is not theirs. So when their time comes to want to tell us something, they will burn it.

Like the adults, students resorted to uncooperative methods of problem-solving when denied rights and opportunities to contribute to their own development. Age and experience, thus, accentuated domination and denial of rights of agency. As shown in Chapter 4, a contributing factor to violent and non-cooperative ways of problem-solving was the inability of SMCs/BOGs/PTAs, technocrats and donors to act in transparent and accountable ways. The use of age and experience as determinants of success endorsed the view that development is something done for people, not something people originate. Such practice is dictatorship of development (Ife, 2002) and tyranny of participation (Chambers, 2005; Cook and Kothari, 2001).

It is dictatorship because it treated SMCs/BOGs/PTA, disadvantaged people and students as recipients of development designed elsewhere. Such practice arose from the inability of the donor and technocrats to treat these people as equal stakeholders. This practice led Sundukia, a parent/BOG member, to caution:

> Although in most areas these people have been taken like recipients. I think it is a major challenge. I think that is somewhere we are failing because, like when you look at secondary schools, really we are dealing with mature people. And even in primary school, I don't think there is anybody who is so young and small that he has no opinion of what is happening to his/her own life. If the pupils/students were also given room to influence what happens to them, I think they can also influence in some way … the advisory services given by the ministry [or] government.

Considering that the BEIP was in its first phase at the time of this research, SMCs/BOGs/PTAs emphasized the need to actively involve students in the manual activities generated through the construction plans and in decision-making. Failure to involve students decreased benefits of ownership and negated their own responsibility where they engaged in destroying school property at times causing death. These violent and irresponsible practices increased educational costs to the parents who had to rebuild the schools. By enacting the BEIP through an aid delivery system the government risked shirking its own responsibilities through participation and increasing costs and taxes to the already overtaxed and overburdened members of society.

The failure to engage students negated their own responsibility, threatened sustainable development, and risked perpetuating dependency on external aid. Imperatively, to increase upward accountability to the sources of the BEIP funds and downward responsibility to the recipients required the involvement of the disadvantaged people and the students in enacting and implementing the BEIP. Engaging students and communities in enacting, not just implementing, the BEIP would increase their responsibility. However, this demanded an attitudinal change on the part of technocrats, political elites and school administrators. Key elements of the change process included downward responsibility to the students and communities, collaboration, consensus-building in decision-making processes and an educational (or consciousness-raising) process that allowed for learning in tandem.

Nonetheless, as promulgated in the BEIP, much of this discourse was one of domination. To enable development for students, even when they are told or informed about the sources of funds and how their parents are involved is not enough. Although the BEIP had some spaces for students to participate, cultural perspectives of age and experience negated the benefits of collaboration, consensus-building and learning in tandem because students were excluded from most of the implementation and decision-making processes. Such exclusion arose from the view that children have no rights and that their parents and SMCs/BOGs/PTAs know what is in their best interests. It is conventional practice in African contexts to deny students their rights to contribute towards their own development through views such as " … they are still children" (Sundukia). Yet, this negated benefits of empowerment and social change to them.

While acknowledging the worth of government and SMCs'/BOGs'/PTAs' responsibility towards students, the way the constitution promulgated rights of students limited their participation in development because they were considered children whose voices should be heard through their parents until they attained 18 years of age. Viewed through cultural norms that encourage collective responsibility toward children, in the target schools voices of students were to be heard through teachers. Nonetheless, to treat students as recipients of development because they have no citizenship rights is to negate their own responsibility and that of the duty-bearers (government and parents/communities). Again, the social and political embeddedness of rights as stated in the constitution, their implications to government and parents' responsibilities (as duty-bearers and promoters of children's rights), the cultural perspective of children and the complexity of participation limited effective understanding of how age and experience reinforced inequalities. One teacher said:

> If we want these young people to be an integral part in our society, and to involve them actively in participation, then we should be able to move together so that they are able to see our strength and that is the intention. But also let them see also our weaknesses ... I mean it is an internalised process that a child is going to ask you what they are not sure of or whatever you have done or said. And you will not turn to him and say that is not your responsibility. Why are you asking? When were you born? Whose is older? Yet we are talking of ideas. (Shirikiano)

Devaluing the rights and ideas of the youth, children and young adults was well entrenched in the settings studied as part of this book. A key finding is that excluding students from decisions that impacted on their lives denied them opportunities to develop critical life skills such as truthfulness, openness, dialogue, assertiveness, teamwork and cooperative ways of problem-solving. Arguably a lack of these skills led them to resort to violence and non-peaceful acts of problem solving. To effect social changes required 'unlearning' of the dehumanizing practices that treated students, SMCs/BOGs/PTAs and the disadvantaged people as recipients rather than creators of development. The unlearning process demands reorientation of mindsets and realization of political and human rights. The social and political embeddedness of these rights imputes a responsibility on the government and leaders to account for their actions to the citizenry by enforcing individual and collective rights (including rights of access to economic activity and source of income). Otherwise, the denial of 'spaces' to influence development on the grounds of age and experience created a dependent society where children relied on their parents for solutions and communities and government on donors for aid and strategies. These risked entrenching perpetual dependency and inequalities of poverty.

Poverty

The way technocrats promulgated the structural and rights-based approaches suggests a systemic approach to the question of poverty, which according to Ife (2002) should apply at the level of policy, practice and analysis. At the policy level, poverty was taken to mean the "inadequacy of income … deprivation of basic needs and rights, … lack of access to productive assets [land, employment and] social infrastructure and markets" as a result of natural disasters and an inequitable distribution of wealth, goods and services (GOK, 2002, p. 6). Arising from this understanding, the BEIP aimed to increase children's access to the rights of education and communities' participation in the management and governance of education.

Thus, the BEIP was meant to create an enabling environment through which to reduce poverty and achieve sustainable human development in the long term. Towards this outcome, technocrats chose to implement the BEIP in areas that were perceived to have high levels of poverty. It was believed that poverty had led these disadvantaged communities to be unable to meaningfully participate and benefit from educational development compared to other communities. The value these communities attached to education and development led them to readily accept the BEIP and the idea of people-centered and participatory approaches. They understood their needs and how the project was likely to impact on their livelihoods. However, as an *effect*, poverty significantly limited the agency of disadvantaged people even after they were provided with structures and opportunities to participate. Antoinette stated:

> There are some areas that are really hit by poverty, such that however much you expect the community to participate, it is hard hit. Some of them are not able to … come to help in manual labour. They have no strength. They have no

> food. Others have nothing to offer ... They have no sand to give. In some areas the project may take up at a slower phase. In other areas which are endowed economically, I realized that the project tended to blossom quite fast ... There are variations [however].

This idea that poverty determined the extent and types of participation and also the pace of development is critical to our understanding of the impact of the BEIP. There were marked differences in the way poverty affected the participation of agriculturalists, agro-pastoralists and pastoralist communities. Generally, poverty motivated most communities to participate, but structural disadvantages limited the extent to which this participation delivered tangible benefits. The finding that poverty either motivated some communities to participate while discouraging others questions the validity of the technical approach used to enact participation. The technical approach 'universalized' or 'generalized' participatory approaches and discouraged participation of some disadvantaged individuals and groups. As noted, the differences in socio-economic and environmental set-ups emphasize the need for contextualized development approaches.

When policies and approaches are contextualized, benefits of ownership and cost-effectiveness are more likely to be increased. This is because despite disadvantaged people's willingness to participate, natural and environmental conditions limited their participation and variedly increased the costs of the BEIP. Contrary to the view that participation increased efficiency and cost-effectiveness, environmental factors limited alternative choices of the disadvantaged people and those of technocrats:

> … in areas where they have environmental challenges, we had no choice. The work that has been done is not as much as in other areas … We have those discrepancies … in hardship areas, they have spent more in getting material compared to those who get materials from close proximities. (Carla)

Considering the BEIP's focus on empowerment, it could reasonably be expected that in implementing participation, technocrats paid attention to these socio-economic and environmental conditions. Despite awareness that these communities lacked access to sources of income and were unable to meaningfully exploit their environments, technocrats prescribed that all disadvantaged communities must participate both in human and physical capital to enhance ownership of the BEIP. This was coated in the language of 'voluntary participation'. Arguably, it offered the disadvantaged people 'freedom' to participate based on their material abilities. Such a prescription of participation, as a technical panacea to poverty, negated the contextual differences. It risked accentuating poverty and regional inequalities. Carla said:

> Some regions are rich. The parents are willing to contribute. They are putting great effort. Some communities are very poor. They cannot contribute anything.

> Even if they came to work it is like they should be given some food … because they spent all the time looking for food. So when we … engage them in … school activity, [they ask]. At the end of the day can we get some food to take home for our children? In these hardship areas, it is a real challenge to tell them to put one or two hours into school activity … They spent a lot of time trying to secure food for the children.

The quality of the 'choices' offered through voluntary participation can, thus, be questioned. Choosing not to participate in the BEIP to earn an income to feed one's children is an act of responsibility and empowerment on the part of disadvantaged communities. It concurs with Ife's (2002) contention that people participate when they know their actions will make a difference. However, to assume that the richer communities were more willing and able to participate than the poorer communities is to blame the less endowed communities for their own conditions. Indeed, structural disadvantages and disadvantaged people's relationships with the structures that governed their lives limited their ability to participate. For example, as said earlier, technocrats required disadvantaged communities to voluntarily contribute their time, materials, labour and ideas but not money (which they did not have). This redefinition of poverty as 'lack of money', on the flipside, meant that disadvantaged people were rich in human capital and could afford time to attend to the BEIP activities and processes. However, "most of us do not have that time because we want to go out and earn some income to feed our families" (Jamal). These socio-economic diversities and complexities of participation suggest that poverty is more than a lack of monetary capital which can be addressed through aid. Thus, the technocrat view of poverty was limited and risked accentuating regional and class inequalities without necessarily addressing the root causes of poverty. Mapatano said:

> Concerning this issue of poverty, this issue of voluntary work sometimes is very complicated because those people we want to come for the meetings, they have families and they are not paid. So they have to run for their own survival. That is why they have limited time to be at the site … Poverty is a key hindrance.

These outcomes spell out the contradictions of participation and the use of local resources, materials, knowledge and skills to spread aid benefits to these communities. The use of local potentials was taken to mean selling goods and services. Important as this is, the paradox is that the processes of commodification ensured that only technocrats, technical experts, builders and contractors were paid. In enacting the BEIP, technocrats assumed that parents and SMCs/BOGs/PTAs should provide 'voluntary' or 'free' labour, ideas and time to increase cost-effectiveness, ownership and responsibility. These perspectives of poverty, allowed for impositions of conditions of participation that poor communities were unable to satisfy. Reuben said:

> In relation to the issue of poverty, there are some donors who have a condition, 10% contribution from the community. This is where the problem is with OPEC. The contribution in terms of participation cannot be raised. If they come to dig trenches and they go home with nothing, the following day they do not come.

This idea of remunerating technocrats and technical experts partly inhibited participation. The failure to involve the disadvantaged people in designing the BEIP led disadvantaged people to believe that technocrats used their poverty conditions to justify their own interests in aid and donor interests in markets. For this reason, disadvantaged people also asked to be 'paid' a token amount to be motivated to participate. While the communities' attempt to ask for payment could be justified on moral grounds, as said before, these ways of addressing poverty are tokenistic and were perceived to entrench corruption. The data described corruption in terms of the failure by the government, donors, SMCs/BOGs/PTAs and other change agents to accomplish obligations as claimed in the BEIP or to enact policies to justify technocrats and donor interests in aid and markets. As noted in the data, although the BEIP was justified on grounds of poverty, ASALs communities believed that they were not poor. Rather, technocrats and donors 'perceived' them as poor to market goods and services from the more endowed areas. Dororomo, a district committee member, avowed:

> I believe our people are not poor ... They are poor because of the perception. They are only perceived to be poor and because everybody is saying that they are poor, they have accepted that status. Why am I saying this? The resources we have in most of these areas are untapped. Our local resources are untapped 100 times. They are merely underutilized. Why do I say this? These projects come with millions of shillings. I wish to take the OPEC project in Mandera … Much of the shillings [go] back to Nairobi because of the cement and the sand, mostly the cement, not the sand. There is that locally available building material in our local communities.

This finding that the BEIP facilitated the exploitation of disadvantaged people suggests that participation in aid projects is out of step with Ife's (2002) view that development interventions should increase the power of disadvantaged people over personal and life chances, power over economic activity, and power of reproduction. Thus, to provide solutions to poverty through an aid delivery system is to disempower disadvantaged people and entrench dependency:

> Disempowerment [is] when you lose your sense of dignity, capacity … what I am seeing in schools, if they [government, donors] continue supporting them like this without sense of responsibility, when they withdraw, some of the stakeholders and especially the parents … will have lost their sense of responsibility. The dignity to be responsible parents … [T]hey look at themselves like other people and somebody should come and help. That is the worst benefit of the OPEC

> project. The moment you get disempowered, you don't feel like you can do anything for yourselves. You no longer feel in control and you cannot even have a vision of the kind of change. I think that is the greatest loss an individual/ community can get. (Sundukia)

To the extent that the BEIP left structural poverty and class inequalities unaltered, encouraged dependency and negated responsibilities, its contribution to development is unsustainable. This issue is considered more fully in the following section.

Sustainability

This section builds on the view that aid development, as found in the BEIP, is unsustainable. Documents (GOK, 2002b, 2003b) attest that the BEIP aimed to impact upon sustainable development through increasing participation and partnerships. As said before, participation in needs assessment was meant to empower and enable SMCs/BOGs/PTAs to commit on their behalf and on behalf of the broader community of parents towards sustaining the BEIP. As part of this commitment, SMCs/BOGs/PTAs were expected to initiate income generating activities such as:

> ... open[ing] up these institution(s) to the community such that when ... learning is not going on within the school, ... any one can hire the school facilities and use them for their own personal benefits. For instance, if you hire ... the school compound and buildings for a wedding ceremony or family gathering, ... then you pay ... Those funds are meant to sustain the project but on condition that [the users] will not affect the future use of that facility. So, there are others who will not give anything, they may not even be directly involved in the implementation but they can become users and indirectly participate in sustaining the project through what they shall give to the school when they hire. (Antoinette)

Other income generating activities reported in the data were planting trees, vegetables/farm-crops and dairy farming. Fundraising through CBOs/NGOs, government grants and private entrepreneurs was also identified. Thus, sustainability was construed in terms of maintaining the physical facilities, built through the BEIP. Participation and partnerships were, thus, *methods* of sustainability. The data attest that participation increased the awareness of SMCs/BOGs/PTAs and parents and enabled disadvantaged communities to sustain the BEIP. Bushie stated:

> People have learned to sustain projects with the resources they are generating from those projects ... Depending on how capacity-building has been done to the various stakeholders, sustainability can be achieved ... If the component of capacity-building has not been well articulated in the entire process of the

> project, nothing much will be achieved. People will continue waiting for donors to assist them.

Again, as a result of the consciousness-raising processes "communities have learned to ask for returns for what they give. [For example] land, they expect what is constructed there to benefit them" (Bushie). Despite enhancing awareness of the BEIP to the broader communities, the data are pessimistic about the extent to which advocacy, capacity-building and participation empowered the disadvantaged people to actually sustain the BEIP and initiate future similar development projects.

Moreover, despite awareness that sustainability required balancing the technical and moral components of development, the focus on structural reforms rather than transformative reforms meant that the disadvantaged people did not have the technical and resource abilities required to sustain the BEIP. This imbalanced implementation has been shown earlier in the use of questionnaires and pilot testing techniques, phased implementation, and the withdrawal of the BEIP from some schools to accommodate donor interests. Such practices posed major challenges to sustainability. For example, the withdrawal of the BEIP from primary schools in Coast Province, meant that aid development is unsustainable.

The withdrawal led SMCs/BOGs/PTAs to believe that projects that are implemented in phases or where funds are disbursed in instalments are unsustainable, since such an approach gave emphasis to the interests of donors/technocrats and political elites. Where political elites were not re-elected to power, donor-states withdrew funding based on claims of poor governance or unpopular policies. Mapatano stated:

> … we are just being used by westerners … These [donors] think that we are corrupt, we cannot use money, and they think we have all the time to do what they want. If you go to Europe now, they have every minute planned for something. They do not waste time. For us they give us [instalments of] five hundred thousand. They think we are going to stay with that five hundred thousand for one year before they give us another stage. *So they just spoil us more.* Instead of giving us the whole sum of money we tender, we get the material and buy [services] and finish up. A project for three years takes five to seven years, only for 2.1 million Kenya shillings [or $ US 26,923]. They just give little by little, what is that?

The BEIP funds were directly credited into the school accounts for SMCs/BOGs/PTAs to access directly. These practices generated from financial management roles limited capacities of these communities to sustain the BEIP, gave donors power to 'shift goal posts' as they desired, and risked delegitimizing the government because disadvantaged people lost confidence in the GOK commitment towards addressing their human developmental challenges in emancipatory and sustainable ways:

> … in this project they propose this, they propose that and then their follow-up is not proper. Why? People come to stop because the assistance has stopped. There is no monitoring, there is no follow-up. They sponsored no particular people visiting [communities], living with them, knowing their problems. Just to pump money in a particular district and then you say you have assisted people is not enough. People [development facilitators] need to go and live with the people and see the situation and observe how the programme itself has been taken by the people on the ground … When, they [donors] just sent money, this money can disappear in between before it reaches the destination … I think a project like OPEC needs close supervision, monitoring, guidance on how they want this programme to go and then the sustainability of the programme should have been well set out. When such things are left out, the programme dies naturally when donors withdraw. (Nashika)

The data with SMCs/BOGs/PTAs show that the changes impacted through the BEIP were 'tokenistic' and risked creating social classes of 'haves' and 'have-nots' or accentuating inequalities. Ruaikei, a parent/PTA member, confirmed:

> … you sacrifice at long run when the donor … of the project has gone away, your hand is tied … There is not an inch you can move … You don't have the funds and your people have not been adequately sensitized on how to sustain this particular project. This person employed by the government will just carry on the project supervision because he has the money at hand. But, immediately the donor goes then you [community] have no where to go. There is nothing which has been left for this project to run for some time while the sustainability itself has to be organized. So you are left hanging … and then the project dies slowly. Halafu [and then] … these projects are like drops of water in a desert. Like they will always say this project is piloted in two divisions in a certain district. Like in my case these children had nothing completely. It is a boarding school, the dormitory or classrooms may be there but they are empty … the communities are poor … Whom do you sell the water to … they cannot afford … You incapacitate people distributing these. Now all the schools are needy … Whom do you give and whom do you leave? And, when you give about ten pieces of furniture in each school … and then you go … You know, these [facilities] are floating because they will only be used by few … You even create inequalities … You now wonder, is it better to have some sleeping on the bed, others on the floor or some sitting on the desks and others on the floor.

These concerns meant that the BEIP was unsustainable and the furniture, classrooms, water and sanitation facilities it provided could not meet the needs of the school populations and the neighbouring communities that greatly lacked access to these rights. This way, the BEIP stifled local innovation and motivation to participate while entrenching dependency. Hamish said:

> It is good that the OPEC project is entirely giving the whole lot of money for the whole project to take off and be completed. Otherwise, if there could be a certain contribution of a certain percentage from the community, then it could be stored somewhere [savings] because of the political environment/ atmosphere. If there is a certain percentage for the community to contribute towards that project, then of course the community could go to the politician to ask for assistance. This is where now the community will be deceived. That either I will bring four lorries of sand which will delay or I will contribute four thousand blocks which will delay or the donors give the whole amount of money. So there is no need for the community to contribute. We have come across all those [scenarios] ... Sometimes if the community is supposed to pay ... some ... donation, politically these problems come in. There is an element of something being given free now, so the community member should not be [asked to contribute].

A contributing factor to the inability of the BEIP to affect inequalities and sustainability came from structural disadvantages arising from environmental, economic, educational and political factors as shown in the following subsection.

Environmental, Socio-economic and Political Factors

From the perspective of technocrats and SMCs/BOGs/PTAs, sustainability is a significant element of emancipation. However, technocrats' focus on the maintenance of the physical facilities enacted through the BEIP meant that sustainability depended upon the ability of the disadvantaged people to create markets and make profits from the income generating activities they were able to establish. Indeed, the way technocrats promulgated the BEIP and the use of local potential fell well short of satisfying the poverty and developmental needs of disadvantaged communities. Firstly, the approach excluded disadvantaged people from processes that identified and mobilized resources to address their poverty and educational challenges organically. Secondly, the socio-economic institutions and infrastructural systems needed to enable meaningful participation were lacking. The use of representatives, as said before, led to the creation of elite-to-elite networks. Rather than providing permanent solutions, these approaches accentuated class inequalities and further marginalized the disadvantaged. Sundukia contended:

> I think it is a deliberate way to marginalise others ... The whole issue of participation is very closely linked to poverty. The moment this person is denied an opportunity to an education ... take that as exclusion. This is because the issue of information being power, knowledge being power ... You can't get yourself a job. You can't argue out your case clearly in the community and you cannot get feasibility ... It goes up to the national level. You will always [be] excluded as an individual and community and of course poverty becomes a reality.

Far from increasing power over institutions of health and education (Ife, 2002) and, thus, their ability to sustain the BEIP, the education sector was perceived to marginalize and entrench poverty through policies which encouraged admission to universities on the basis of available beds. The criteria used to select primary graduates for admission into secondary schools also satisfied the interests of the elites and the rich. An example is the policy on 'quarter system' which ensures that only 15 per cent of primary school graduates in a district are admitted into national and provincial schools while 85 per cent are enrolled into schools within that district. This criterion encourages elites to enrol their children in high-cost private primary schools which are within provinces where highly academically performing schools (of their choice) are located. Since the poor could barely afford the high costs in both the national and provincial school, these practices and policies led disadvantaged people to believe that poverty is 'created' through denial of rights of access to education, economic-activity, health and security, and discourses of representation and aid. Dororomo said:

> Look we have been given a name tag. This name, 'the ASAL' and then there is 'hardship area'. All these are labels that we are not ... As an individual, I am not comfortable with these labels. Because it is segregating us, it is dividing us. It is separating us from the rest of Kenyans. Why should this be used if they are assisting us as they claim? I think there is this saying which says that instead of giving somebody fish for lunch, it is better to teach him how to fish. Yes, the area is hardship, but we should use that hardship to positively develop our people. That condition [dry weather] is not available with everybody. There is nobody who gets 24 hours or 12 months ... sunshine like us … Most of these developed countries … do not have 24 months sunshine … They do not have 12 months working period of the year. Half of the year they are covered with snow, they cannot produce as much sun as we are producing. So we have this time, and we have this environment at our disposal which is untapped … [Yet] when there is disease, it is ASAL people, when there is war it is ASAL people, when there is drought, it is ASAL people, when there is floods it is ASAL people … Insecurity, ASAL people, why? I don't think we have more insecurity than ... Nairobians. In these projects, our people should be involved in resources identification and mobilization … for their development … they must be involved in mobilization of their own resources. That is the only time they can own up these projects and the project can be sustained … Most of these development projects run by the ministry like GOK/OPEC lack networking and effective coordination ... That is the way I feel. That is what hinders development.

The GOK failure to develop enabling environments for ASALs to meaningfully exploit their socio-economic and environmental capital limited benefits of sustainability. SMCs'/BOGs'/PTAs' testimonies questioned the logic used to describe these communities as poor and their environments as 'ASALs', 'hardship' and 'insecurity zones'. Such 'labels' meant that the government justified aid through education and other economic development policies and used these to further marginalize pastoralists. Dororomo added:

> Education policies are surely biased … The policies in this country generally are just biased to us the pastoralists. Why? Since late 1979 there was no allowing of marketing of livestock and livestock products. While we have Pyrethrum … Coffee Board of Kenya, Kenya Tea Board, … we don't have livestock production and marketing board in this country. And we are boosting economy … We provide more that 68% of the beef … livestock products to the country. Why should we be seen as a liability … poor, when we are [supplying] all livestock products to this country? So, we are deliberately neglected in these development strategies.

Where educational and developmental structural disadvantages inhibited access to education, participation and sustainability, the BEIP risked entrenching class, poverty, and ethnic and regional inequalities. School communities were expected to provide security to the schools where the BEIP was implemented as a way of cutting the costs of employing security guards. As part of this role, parents were supposed to build access roads, fence the schools and keep vigilance. SMCs/BOGs/PTAs were nonetheless pessimistic about how the disadvantaged communities were meant to sustain these roles in the long term, largely because of the high crime rate in urban slums and hardship areas. They questioned the efficacy of the BEIP to increase security to communities that were not even entitled to such rights in the words:

> From the time of colonial [rule] … we are said to be … violent/harsh. The climate is harsh and the communities are harsh … That is why people in hardship areas are taken to be very violent people … That is why we … had section 2A that is putting us under emergency from colonial time … People who are under emergency have no right to do anything either education or anything. It was only the other day that the section was removed. But, if you go down now to Mandera, Isiolo, somebody's right is violated and nobody will even think of correcting that. This is the way we have been thinking … and we still think that is still there. We have inherited from all that time. The other thing is ignorance and inadequate education … It is … normally said, if you go to a local man in Mandera, Isiolo or Moyale, he will ask you, do you come from Kenya? That means they are not even Kenyans. They don't even associate themselves with Kenyans. It is not ignorance … They believe that those people from down Kenya are Kenyans. The ASALs are neither Kenyans, nomads nor Somalis. They don't even know who they are. They are hanging wherever they are. (Nashika)

To ask the disadvantaged communities to provide security to schools when their own security rights were not structurally guaranteed risked shirking government responsibility towards promoting such rights while limiting sustainability of the

BEIP. As a result of 'perceived' insecurity, schools in the ASALs were neither maintained, nor supervised. SMCs/BOGs/PTAs feared that despite enacting structures and processes to monitor and evaluate the schools built with the BEIP fund, generally ASALs schools were hardly supervised or monitored as frequently as schools in other areas. These schools lacked facilities and performed poorly in national examinations.

Monitoring and evaluation in the BEIP was also ineffective. Technocrats could often not get through to the schools because of harsh weather and rough roads. Such 'neglect' encouraged elites from disadvantaged communities to devalue their own schools and choose to send their children to schools they percieved to be better in other parts of the country. The paradox is that disadvantaged people continued to miss out on education. The failure to integrate the socio-economic needs of ASALs communities in educational policies meant that when their children completed schooling they could neither secure employment nor herd livestock since they missed out in such life skills. This is different to children from agricultural communities who could resort to farming when they missed out on waged employment. This way the education system attempted to detach disadvantaged children from their socio-economic lifestyles and challenge traditional socio-economic systems without providing alternative sources of income. Thus, a parent contended:

> I think the curriculum is creating the gap? Why is the curriculum not serving our culture? Why is the curriculum not meeting our needs and yet it is meeting the needs of other Kenyans? Because of lack of job opportunities, our children are not going back to herding. Now what do they do? They are involved in drugs and drug abuse ... And then the discipline deteriorates. They try even to influence students who are admitted [in schools]. So lack of job opportunity brings a lot of other problems to us ... [including] HIV/AIDS. (Darren)

These findings concur with Sifuna's (2005b) contention that the education system served better the interests of the communities that lived in the areas of high and medium agricultural potential than pastoralists-nomads living in the ASALs. Economically, this means that pastoralist communities must wait for the agricultural communities to farm and market their goods to them. This way of redistributing goods, services and markets makes pastoralists depend on the agriculturally endowed parts of the country. The way the government promulgated economic development and integrated these policies within the education system served agricultural rather than pastoralist communities through the provision of water, roads, electricity and social amenities. This inequitable distribution of economic assets, infrastructure (including educational institutions) and the poor integration of disadvantaged people's needs in the education system arose from unclear policies and wrong priorities based on the personal interests of political elites, donors and technocrats rather than community needs. Sundukia confirmed:

> Look at the way the [government] has shared the tarmac roads, I don't think there was any merit apart from who is who. In a country where development is linked to personalities other than clear policies, then automatically that means that everybody who has been excluded from education cannot climb up to the top, cannot influence policies, and the moment you cannot influence policies, that means exclusion and once you are excluded then poverty is with you. And it is not only with you for days, it can even be institutionalised. This is because if you are poor and you cannot take your child to school, you bring up another poor person … With time you realize, regions, communities, families are completely excluded.

SMCs/BOGs/PTAs argued that the education system encouraged competition in examinations. Such competition meant that the education system did not serve the socio-economic and employment needs of either ASALs or of other communities, even after achieving good grades of C+ and above. These ways of maintaining a balance in the status quo denied ASALs access to the right of economic activity and employment while forcing youth who have attained academic achievements to continue to 'depend' on their parents for their livelihood. Thus, the education system through the BEIP risked entrenching ethnic, regional and class inequalities at the micro, meso and macro levels of development without addressing the root causes of poverty.

To the extent that the education system encouraged competition without necessarily addressing poverty, education and employment needs of disadvantaged people, it is right to state that cooperative rather than competitive ways of problem-solving are likely to increase the benefits of empowerment and transformation to the hitherto excluded. Thus, as Ife (2002) contends, participatory development interventions must challenge the competitive ethic that informs most education systems with a view to increasing the power of disadvantaged people to control their own future. Decrying the education system, Nashika, a teacher, suggested:

> Education should enable one to use the skills and the knowledge he has attained to exploit the environment [to] develop himself and sustain his parents, not to come back to depend on them … There has to be a mechanism by the government or the civil society, whoever is concerned with development of humanity in the republic. That mechanism must aim at improving the life skills of that individual. If you are trained to become a teacher or farmer you must have that speciality inbuilt in you to ensure that you … improve your participation. What is lacking is that proper mechanism … The goal of education is to develop one to live a happy and quality life in society … I mean [being] able to provide for his family adequately.

These perspectives were neither meant to devalue education nor the benefits of aid projects. Instead, they emphasized the need for the GOK to address the economic, socio-political and environmental factors which inhibit disadvantaged people from controlling their own development. They stressed a focus on contextualized participatory development strategies based on environmental and socio-cultural knowledge and the need for transformation as expounded in the concluding statement.

Conclusion

This chapter has shown that bureaucracy, environmental, cultural and socio-economic factors inhibited the long-term engagement of disadvantaged communities with the BEIP. Their poverty, lack of information, time, materials and low awareness levels suggest that unless these structural disadvantages are addressed emancipation and sustainable development are unlikely. To the extent that the BEIP left existing inequalities and social exclusion unaltered, or appeared to accentuate these, a complete overhaul of the physical and social infrastructural conditions of the schools and neighbourhoods and a rethinking of the competitive principles of education and participatory development are desirable. Most importantly, the data suggest that the education system must scale-up empowerment efforts with a focus on transformation, not just structural reforms enacted through an aid delivery system. To succeed, such a project should seek to bring services closer to the disadvantaged people through home-grown solutions (not those borrowed off-shore) as these are not only sensitive to the conditions these communities suffer but are more likely to be sustainable.

The next chapter concludes the book by summarizing the main findings and exploring the implication to participatory development policy, theory and practice and the possibilities for future research.

Chapter 7
Conclusion: Implications to Participatory Development Policy, Theory and Practice

Introduction

This chapter restates the central aim of the book and summarizes the main findings. It also considers how the research findings implicate contemporary debates on mainstream participatory development, Ife's (2002) model for community development and modernization, dependency, alternative development and post-development theories.

The central argument of the book is that although participation and collaboration in the BEIP have, to a limited extent, enhanced the teaching and learning environments of the target schools and increased awareness of rights to the disadvantaged people, accountability has remained top-down. These top-down approaches have contributed to social exclusion and continued marginalization of disadvantaged people. For these reasons, the book argues that emancipation and sustainable development are more likely to emerge through interventions that increase participatory practices, that entail government partnerships with civil society and local communities, which promote structures and discourses of citizenship and rights, and where the grassroots is the locale for change.

Central Aim of the Book

The purpose of this research was to critically examine the efficacy of mainstream aid programmes that embrace people-centered, participatory approaches and government partnerships with donors, civil society and local communities to effect benefits of empowerment and social change to disadvantaged people. It also aimed to utilize structural and poststructural perspectives to critically assess the 'fit' between policy, practice and theory of participatory development and its relationships with participatory democracy, use the knowledge obtained from the perspectives of those directly involved in the BEIP to appraise Ife's (2002) approach to community development, and illuminate theoretical debates that are ongoing in development. It answered four main questions.

The first question looked into the way the BEIP objectives and management structure established the context and policy for participation, partnership, empowerment and sustainable (or balanced and holistic) development. It also considered the selection criteria used to ensure the BEIP impacted on disadvantages of poverty, gender and culture. This book also explored the principles guiding the formation and functions of the management structures and their impact on participatory democracy.

The second question assessed the approaches, principles and impact of the BEIP partnerships. It aimed to ascertain whether there is a 'level ground' on which the government, donor, civil societies and disadvantaged people participated as equal partners, the extent to which social networks emerged across social class and their durability and impact on cooperation and accountability.

The third question examined the extent to which the process and outcomes of participation within the BEIP adhered to principles of change from below. It interrogated levels of involvement by the different actors in the processes of planning, implementation and monitoring and evaluation. Special reference was made to aims, meanings, methods and principles of decision-making, consciousness-raising, community-resource mobilization and the use of local potential. Any opportunities and challenges offered for the participation of disadvantaged people were also highlighted.

The fourth question critically examined the extent to which the management structure, partnerships and participation challenged dominant discourses and structural disadvantages based on bureaucracy, culture, gender, age, poverty and broader socio-economic, environmental and political factors. The themes of holistic development were here once again reflected upon to gauge the potential for the BEIP to effect empowerment, sustainable development and social change to disadvantaged people.

Summary of Main Findings

As Ife (2002) suggests, this research has shown that participatory development is more likely to address the structural and rights challenges of disadvantaged people when a holistic approach to development is used. While development policies may have a strong focus on holistic and balanced development, there existed significant 'disconnects' between the stated policies and the practices generated through the BEIP. This disconnect confirms that despite awareness of the need to balance between structural and rights (or technical and moral) components and an understanding of development as a process, technocrats are usually engaged in processes of translation, and the consequences of their policies and decisions may not be intentional (Williams, 2004). It also attests that translating policies into practices of balanced mainstream participatory development is neither value-free (Ife, 2002), nor is it without structural challenges to the technocrats and disadvantaged people.

As shown in this research, mainstream participatory development continues to appropriate bureaucratically organized management structures. This was founded upon the view that the senior bureaucrats are most able to make feasible decisions and policies and promote relationships based on partnerships and participation.

Arising from this view, participatory development is implemented through 'invited' representatives – these are not democratically elected and their management roles are not legitimated through processes of consultation or other forms of democracy. For this reason, representation is a significant inhibitor to participatory democracy because it legitimated bureaucracy and the related undemocratic practices.

Participatory development within bureaucratic contexts can, thus, appropriate discourses of representation, participation, partnerships and empowerment to exclude the very disadvantaged people from decision-making and policies that affect their lives. Despite decentralization of functions and services to lower tiers of the management structure, the retention of decision-making authority with the central office led to the emergence of participation and collaboration through representatives as a form of new centralism. That is to say, where participation is promoted through invited rather than democratically elected representatives (and often undertaken at the discretion of elites), participatory development denies disadvantaged people the rights of participation, and excludes them from decisions and results in elite-to-elite networks.

This practice, which has confirmed participatory development management within an aid delivery system through representatives (Brown, 2004) and technical experts is indeed a form of social exclusion. Besides negating contextual differences, the practice of participatory development denied the disadvantaged people freedom of choice and opportunities to participate fully in the identification, implementation and monitoring and evaluation processes of the BEIP as participatory development protagonists (Burkey, 1993; Chambers, 1997) advocate. Despite having sensibilities of participatory democracy, the practices of participation, partnership, consciousness-raising and consensus-building that were employed in the BEIP show that participatory development largely remains a discourse of the powerful about the powerless (Ife, 2002). It can potentially legitimate dissenting voices through such terms as stakeholders and development partners, community and promote new forms of professionalism (Chambers, 1997).

A central finding is that to empower and transform the lives of disadvantaged groups, the process and outcomes of participatory development must be treated not as separate, but rather as a process whose means and outcomes are intractably entwined (Ife, 2002). The data reported in this book showed that separation of ends from means encouraged so-called change agents to privilege technical over moral elements of participatory development. In other words, change agents focused on structures, participation, partnerships, empowerment and sustainability, as methods and/or outcomes, rather than processes in which the disadvantaged people are active, not passive participants. The book argues that such privileging comes from technocrats' and donors' top-down mindsets and the use of representatives, aid assistance and technical experts to seal perceived knowledge and information gaps on the part of disadvantaged people.

These features, together with 'dichotomic thinking' (Pieterse, 2002), blind change agents to effectively understand how the bureaucratic contexts, their own perceptions and approaches enhance social exclusion, inhibit participation and the creation of partnerships and social networks on an equal basis. To the extent that the findings of the present research can be generalized, empowerment, transformation and sustainable development are more likely to emerge with interventions that promote participatory practice, that embrace government partnerships with civil societies and local communities and which promote structures and discourses of rights and citizenship.

Nonetheless, a major challenge for participatory development remains overcoming donor interests in markets and technocrats' interests in aid. Despite participatory development claims to redistribute economic surplus and technical knowledge from the already endowed and powerful (Arnstein, 1971), such interests are critical inhibitors to the use of local knowledge, skills, potential and the realization of organic development (Ife, 2002). The attempt to promote participatory development in the BEIP fell well short of realizing change from below and the integrity of the participatory processes. Despite intentions to implement participatory development as a process, the structural approach adopted led to commodification of participation. Such commodification risked accentuating power and market shifts away from the disadvantaged to the donor and enabling the GOK to shirk its own responsibilities to its disadvantaged people.

Where cooperation and partnerships are premised on the need to relieve the government burden by maximizing benefits of aid through microeconomic principles of redistribution, participatory development risked supplanting structural disadvantages (Botchway, 2001) critical to emancipation and encouraging dependency on external aid (Burkey, 1993). As Klees (2001) contends, far from ensuring partnerships on an equal basis, participatory development in the BEIP strengthened donor monolithic power and exposed the vulnerability of the GOK and its disadvantaged people. The imposition of donor power occurs because donors find it hard to change existing ways of thinking and doing things. For this reason, participatory development partnerships continue to privilege donor interests through conditions established in negotiations that are sealed by way of aid memoirs, agreements and contracts, and which are neither accessible to disadvantaged people, nor to their representatives.

Although technocrats negotiated with OPEC, such negotiations established conditions and policies which limited the choices of the SMCs/BOGs/PTAs, collaborating technical experts from other sectors and disadvantaged people. Contrary to literature that sees little, if any, role for governments in the Third World (or at least present these as passive actors) (Rahnema, 1992) in participatory development, such negotiations recognize that governments do have an active and important role and are indicative of opportunities for change – the donor and the government are willing participants despite absence of 'equal partnerships' in practice. As Chambers (2005) states, governments may appear to be in control of aid development projects when some other power, for example a donor, makes the most critical decisions. Though having the potential to affect partnerships on an equal basis, negotiations are potential mechanisms to rubberstamp donor interests in markets and government interests in aid.

Such interests contributed to the emergence of partnerships on the basis of competitive, rather than cooperative, relationships. To the extent that such competition led to the imposition of decisions confirms that more cooperative approaches are able to address structural disadvantages (Ife, 2002). The problem with aid partnerships underpinned by competitive and market principles is that macro/meso/micro-level power relationships are left unchallenged. That means participatory development partnerships are not just empty spaces. Rather, partnerships are spaces filled with political, economic and social powers. Appropriation of these powers through participatory development encouraged upward rather than downward accountability to the people. To concur with Gregory (2007), this research has shown that there is a very fine line between accountability and responsibility. Contrary to conventional notions, which only loosely use accountability to denote reporting to donors (or sources of funds), responsibility entails processes of integrity, transparency and answerability to the people. Mainstream participatory development will have to engender downward responsibility to remain a relevant empowerment and transformative approach to the disadvantaged people.

It is widely argued that partnerships in participatory development contribute to the establishment of durable relationships and social networks through which disadvantaged people can overcome their poverty challenges (for example, Buch-Hansen, 2002). However, data gathered in this research show that in the case of the BEIP, the aid partnerships were neither durable, nor sustainable. Instead, participatory development partnerships within an aid delivery system can actually accentuate social exclusion and inequalities (Pieterse, 2002), and entrench dependency (Burkey, 1993).

The book argues that to be sustainable, participatory development requires government partnerships with civil society, citizenry and disadvantaged people themselves and structures that promote participatory practice, that promote active citizenship and rights agendas (Hickey, 2002; Ife, 2002). Otherwise, dependence on aid curtails local innovations and foresight and risks entrenching corruption rather than addressing it. Technocrats' tendency to enlist local participation to increase efficiency and cost-effectiveness of aid redefined participation in apolitical terms (contribution and sharing). This necessitates participatory development researchers and practitioners to (re)politicize participation (Williams, 2004) with a view to enabling disadvantaged people to initiate social and political action (Ife, 2002) on their own disadvantages, which indeed are a denial of basic human rights.

The practice of participation in the BEIP confirms that for most of the time mainstream participatory development continues to be enacted in a top-down manner (Ndengwa, 1996). Where disadvantaged people and their representatives participate, the most important decisions are made by technocrats in collaboration with the donor. This failure to involve disadvantaged people in decision-making processes attests to the fact that participation in participatory development practice is largely either tokenistic (Arnstein, 1971) or coerced (Chambers, 2002). Again, these forms of participation are nominal and instrumental (White, 1996). This privileging of functional over transformational modes of participation emerged through the methods of planning and decision-making.

To confirm Brohman's (1996) concerns, many of the participatory development decision-making practices under the BEIP turned out to be forums in which disadvantaged people did not participate. Methods of planning inhibited the full participation of disadvantaged people and their representatives and curtailed their freedom of choice. Similarly, the methods used in advocacy, consciousness-raising and capacity-building treated SMCs/BOGs/PTAs and disadvantaged people as passive recipients of knowledge channelled through lectures/seminars by technocrats and technical experts. Indeed, these forums were seen by elites and others to increase awareness of SMCs/BOGs/PTAs and contributed to processes of learning in tandem.

However, in practice awareness of rights to disadvantaged people was minimal because they were largely excluded from these educational forums. Amongst other things, these practices confirm that planners and policy-makers lack theoretical knowledge on participation and contextual understanding of the people whose development they plan for (Chambers, 1974, 1983). For these reasons, change agents like disadvantaged people require more empowerment through processes of learning in tandem, immersion, broad political and civic education programmes to affect social and political action than is currently promulgated in mainstream participatory development.

As other research has shown (Cornwall, 2003), processes of community-resources mobilization in the BEIP succumbed to neoliberalism. As a result, the root causes of poverty were not addressed. Unless these structural inhibitors are replaced with enabling environments, technical changes at an individual or organizational level will achieve only limited results. Despite a willingness to participate, these factors inhibited disadvantaged people's meaningful participation and obscured empowerment and transformation. Community-resource mobilization presents an opportunity for the GOK, CBOs and disadvantaged people to jointly address structural and rights deficits. Through this approach, the BEIP contributed to the development of physical facilities in the target schools. This is no mean contribution towards enhanced access to the rights of participation and education.

As Ife (2002) contends, participation in the BEIP also provided disadvantaged people with an opportunity to share experiences of disadvantage and visions of how to combat their inhibiting factors. However, unless disadvantaged people are enabled to undertake social and political actions on their own disadvantages, participatory development will continue to be a discourse of the powerful about the powerless. The research reported in the book shows that the government must provide an enabling environment to promote the rule of development by the people (Ife, 2002) in aid projects. It must avoid manageristic systems (Mulenga, 1999) that are too costly, which bring few benefits to the disadvantaged people and which rather than build communities (Ife, 2002), accentuate social exclusion and domination.

As shown in this research, aid assistance in the BEIP meant that participatory development encouraged the observance and construction of disadvantaged people's development from the vantage point of outsiders (Spivak, 1985). When observed from the vantage point of technocrats, disadvantaged people's lifestyles and their cultures are idealized and targeted as problems for projects to respond to without unpacking such cultures or seeing them as products of internalized power relationships (Williams, 2004). For example, technocrats' perceptions about poverty 'deprived' disadvantaged people of their humanity, while the equation of experience with wisdom cemented with cultural practices to deny women, youth/students and children rights and leaderships positions (and the related roles and responsibilities). These practices precipitate cultural-ethnic animosity, conflicts of interests and violence. The implications of the present research suggest that participatory development will have to appropriate the more cooperative, not competitive; peaceful and non-violent ways of problem-solving; and value culture as a system of knowledge-power from which efforts towards emancipation and sustainable development can benefit.

Participatory development must also appropriate affirmative actions to overcome the deleterious effects of gender, culture, age and poverty and to ensure that pluralist and elitist views do not further marginalize the hitherto excluded from development processes. An area of focus should be instituting affirmative action policies that seek to change technocrats' view that disadvantaged people must compete with others as equals as this negated their meaningful participation. As Pieterse (2002) contends, approaches that enhanced competition in the BEIP were perceived to accentuate micro-meso-macro level inequalities and allowed technocrats-donors to 'experiment' on them with unproven policies (Visvanathan, 1988). Such practices must be replaced with more radical structural and poststructural perspectives that find a balance between experiential-cultural and universal knowledge, civic-citizenship[1] and human rights agendas to avoid the deleterious effects of ethnicity and culture and the imposition of global practices.

A core finding on rights is that participatory development in the BEIP tends to emphasize individual over collective rights of ethnic formations. To the extent that the findings of the present research can be generalized to other settings, to overcome the individualizing effects that result from this liberal approach, participatory development focus should not be to celebrate culture and citizenship indiscriminately. It must identify the salient features that support individual and collective rights and sustainable livelihoods in balanced ways that do not exclude

1 As promulgated in Kenya, the concept of citizenship assumes unitary meanings that negate cultural diversity or treat tribes and cultures as negative terms within democracy (Murunga and Nasong'o, 2007). However, as an element of democracy, citizenship must celebrate diversity (ethnic tribes and their cultures, gender, special needs groups) to bring out the uniqueness and value of each.

disadvantaged people. The need for a broader educational programme than was realized in the BEIP is a critical element for increasing awareness and political and social action. Participatory development interventions should, thus, aim to provide such education through multi-sectoral approaches (including formal schooling, and informal education programmes focusing on health, water, education, agriculture, environment and wildlife sectors). This means that media organizations and civil societies have critical roles to play in promulgating such education. These 'alternatives' (Escobar, 1996) are more likely to enable participatory development practitioners to shift the focus towards communities and villages (Friedmann, 1992) where organic development is more likely to emerge.

A critical limitation to the realization of organic development as shown in the BEIP is that technocrats hardly understood the realities of disadvantaged people. To increase such understanding, the book argues that participatory development must integrate immersion programmes for technocrats to live among the disadvantaged communities (ethnic tribes) when implementing aid interventions. To concur with Burkey (1993), these immersion programmes are intended to enable technocrats to learn about disadvantaged people's ways of thinking and development realities but also increase their understanding of the way culture, tribal-territorial boundaries, which manifest the realities of these people affected participatory development practice in aid projects. Lack of such immersion programmes in the BEIP meant that disadvantaged communities considered technocrats' policies and decisions less sensitive to their realities and thereby unfeasible.

To increase such sensitivity and enable disadvantaged people to initiate social and political action on their inhibiting factors, change agents must rethink their ways of encouraging development and adopt a 360° reorientation – they must change together with the disadvantaged people. Change agents will have to unlearn much of their bureaucratic, capitalistic, technocratic and elitist socializations which make them treat disadvantaged people as 'inferiors' and unable to know what is in their best interest. Disadvantaged people will have to unlearn much of their apolitical and cultural socializations that make them devalue their own potential and humanity. If participatory development is ever to achieve emancipation and sustainable development, human development must retain a central place in policy and planning, education and consciousness-raising, and social and political action (Ife, 2002). For these reasons a more *radical* approach to mainstream participatory development, which originates with disadvantaged people (Friedmann, 1992; Ife, 2002; Pretty, 1995), is both essential and desirable.

In this case, the efficacy of participatory development and its approaches will not be measured by the number of physical facilities and amount of money aid agencies have contributed. Rather it is in terms of how its efforts break ties of dependency (Burkey, 1993) and contribute sustainable livelihoods to the generality of the populace, which can never be attained unless disadvantaged communities (women, youth, and children) are fully engaged in development. Thus, an imbalanced implementation of participatory development through structural rather than change focused reforms limits benefits of ownership, empowerment and sustainability in aid interventions. Implications of these findings to modernization, dependency, alternative development and post-development theories will now be given.

Implication to Development Theories

A central implication of this book's findings to modernization theory resonates with the dual societies (Frank, 1969), neoliberalism and technocratic planning and managerialism that pervade donor-to-government-led development (Willis, 2005) (such as those identified with World Bank and OPEC). Like Frank's dual society, this research has revealed that capitalist and subsistence economies co-exist in the theory and practice of mainstream participatory development among disadvantaged people. However, the findings did not support Frank's view that the capitalist economy dominates and makes the subsistence economy play a secondary role in development.

While in theory the capitalist economy masquerades as the dominant socio-economic order, in practice the majority of the disadvantaged people continue to primarily depend on subsistence economies for their livelihoods. As shown in this research, economic growth structures have neither solved the poverty and educational challenges of the disadvantaged people, nor totally erased the significance of their traditional institutions to their wellbeing. Indeed, the attempt to address the poverty and educational challenges of disadvantaged people through representatives, aid assistance and technical expertise cemented into neoliberalism, contributed to social exclusion and enhanced dependency on elites and donors.

These findings conform to structural, dependency, postcolonial (Bhabha, 1994; Cardoso and Faletto, 1979; Spivak, 1985), alternative development (Burkey, 1993; Chambers, 1997; Friedmann, 1992; Martinussen, 1997) and post-development (Escobar, 1996; Sachs, 1992) ideals. As is the case with structuralists, this research revealed that the GOK has a significant role to play in planning and human development. However, the proclivity towards capitalist, market, technocratic and bureaucratic ideals, as Manzo (1991) argues, led the donor and technocrats to equate the nation-state with the political subject of development (disadvantaged people). For example, from the perspectives of school communities, technocrats formed partnerships and determined who would participate according to the interests of the Ministry of Education and the donor. Such an equation of technocrats' interests with those of disadvantaged people concealed contextual differences and histories of the disadvantaged people. It blinded technocrats from effectively understanding how their nationalist proclivities redefined culture within tribal and national confines and reproduced culture beyond these territorial boundaries (Bhabha, 1995; Spivak, 1985). Structural disadvantages relating to culture, gender, poverty, age, bureaucracy and other economic, environmental, political factors are given a secondary role in participatory development.

These features mattered inasmuch as they either limited or enhanced participation in aid development (the BEIP). And yet, in enacting structures of the BEIP, technocrats and the donor did not consider how their own practices inhibited the disadvantaged people from enacting their own development. This way, the practices of partnerships and participation through representation, aid assistance and technical expertise either negated contextual differences or contributed to 'universalizing' effects to maintain balance and legitimate the status quo. The emergence of new centralism, new professionalism, and competitive rather than cooperative relationships, violent rather than peaceful and non-violent means of problem-solving, despite the use of alternatives of consensus-building and participatory approaches attest that the GOK needs to rethink and reorient its vision of mainstream participatory development for disadvantaged people.

A central theme of reorientation comes from the belief that donors, technocrats and civil societies represent, and are able to transform, themselves to speak and act on behalf of the disadvantaged people. In the BEIP, this notion meant that these elites and elite groups can impose their capitalist, bureaucratic, market, elitist, technocratic-managerialist ideals on SMCs/BOGs/PTAs and disadvantaged people. For example, the definition of poverty in purely macro-economic terms as the lack of economic capital, and the 'reduction' of inequality and diversity issues into a discourse of 'fairness' indeed led technocrats to privilege aid assistance over potential of the disadvantaged people whose lives the BEIP sought to improve. To the extent that these ideals led disadvantaged people to devalue their own humanity, knowledge bases and potential, is sufficient ground to argue that to promote human development in a more balanced way, mainstream participatory development must seek to develop the subsistence economies on which the livelihoods of the disadvantaged masses depend. Here, the focus is not on creation of profits and markets. Rather emphasis is on enabling the disadvantaged people access to economic activity (or (un)waged employment and land) to support their basic needs, which is their right.

The findings show that alternative development (Brohman, 1996; Burkey, 1993; Friedmann, 1992; Ife, 2002) ideals have the potential to facilitate this new vision. Nonetheless the tendency towards structuralism and the subsequent consequences of neoliberalism, technical expertise and managerialism mean that the new vision must appropriate and balance between structural and poststructural perspectives to achieve human development. Here, participatory development must not uncritically integrate non-contextual strategies for economic growth or capitalist economies that are self-destructive and which integrate development of disadvantaged people to legitimate the status quo. Rather, the new vision must focus on 'alternatives to development' (Escobar, 1996). Such alternatives may include affirmative action and anti-growth policies based on cooperative (not competitive) and peace and non-violence approaches (Ife, 2002).

Other alternatives include developing the informal and subsistence[2] economies of disadvantaged people which, according to Kipngetich (2001), together support about 94 per cent of disadvantaged people. Considering the scarcity of resources and high levels of poverty, it is unlikely that the GOK will step up to this challenge alone. Thus, a reorientation of participatory development vision is imminent to ensure collective and individual rights matter more than economic progress. The vision on alternatives means that development is not just about reducing poverty through technocratic and micro-economic policies that reduce disadvantaged people to passive recipients of aid and development strategies and decisions designed off-shore. Instead, disadvantaged people must be initiators and creators of knowledge and prime movers of their own development. The challenge for participatory development is to enable disadvantaged people to unlearn the cultural and apolitical socializations which make them devalue their own potential and knowledge bases.

This transformational process, as clearly shown in this research, will have to focus on the grassroots as the locale for change. This does not suggest strategies for 'delinking' from government-led development (Frank, 1969) since the government has a responsibility towards its people. Although state-led development contributes to Eurocentrism, neoliberalism and technocratic planning, delinking from government-led development 'isolates' disadvantaged people from effective engagement with the political, social and economic structures that govern their lives with consequences of ethnocentrism. To empower the disadvantaged people to be more able to manage their relationship with such structures, it is desirable to entrench the values of participation, cooperation and partnerships into the social change processes. These features will not only strengthen governance, responsibility and accountability, but are also critical steps towards realizing people-led development as the ultimate goal for mainstream participatory development. This means that the new mainstream participatory development vision must engender

2 The post-independent government adopted a capitalist approach to economic development. Nonetheless, the capitalist and non-capitalist regimes have co-existed with competing, but reinforcing characteristics. Despite capitalist economies masquerading as the dominant social order in development rhetoric, in practice, subsistence economies are. Kipngetich (2001) explains that the mainstay of the Kenyan economy is agriculture. Agriculture employs 70 per cent of the workforce but contributes only one third of the GDP. There are varied interpretations to this low turnover. One assumes that rural life is dependent on subsistence farming and on income from urban dwellers, who constitute 25 per cent of the population. This means that there is a high dependency ratio. 47 per cent of Kenya's GDP is concentrated in Nairobi, which makes Kenya to have the second worst income disparity in the World after Brazil. In terms of population, the structure of economy in 2001 was: 6 per cent formal, 21 per cent informal and 73 per cent subsistence. It is currently estimated that there are only 1.95 million wage earners with only few benefiting from commercial agriculture. While subsistence economies are blamed for the high dependency, the capitalist school of thought does not say much about the structural factors shown in this research that inhibit development of subsistence economies.

a coalition that privileges people-led development where donors and technocrats must unlearn treating the nation-state as the subject of development.

The challenge in actualizing this vision is that current participatory development theory assumes that these change agents are capable of transforming themselves through self-reflective methods (Chambers, 1997). Contrary to this view, this research shows that technocrats can (un)consciously enact/malign policies through which the government can perpetuate social exclusion, corruption and the abuse of rights. Even where structures for participation and inclusion have been provided, the failure by disadvantaged people to meaningfully participate because of lack of enabling environments and the weak enforcement of rights by the government means that these features must become central components of mainstream participatory development. This is not to downplay the role of the government. Instead, it is to emphasize the important role of governments in policy, planning and human development.

Despite the said inadequacies of technocratic policy and planning as was promulgated in the BEIP, this research offers clear indications that there are possibilities for change. Such possibilities manifest in technocrats' views on learning in tandem and argue that the policies they enact lead them to trample on the rights and ignorance of the disadvantaged people. These reflective acts by technocrats testify that structures that create opportunities for disadvantaged people to participate in enacting development programmes/projects that affect them are able to increase downward responsibility. Participation is likely to doubly enhance democratic practices where disadvantaged people can claim legitimacy for agency based on citizenship and rights agendas. For these reasons, this research has argued that empowerment of disadvantaged people will largely depend on government partnerships with civil society and local communities (plus youth, women, children and special needs people), participatory practices and structures whose sensibilities for agency link them to citizenship and rights agendas.

Concluding Note

The argument in this book is that mainstream participatory development, as expressed in the BEIP, has not sufficiently addressed the deleterious effects of structural disadvantages caused by culture, gender, ethnicity/racism and poverty. Mainstream participatory development policy, practice and theory attempt to address the structural and rights challenges of disadvantaged people who continue to flounder in their attempts to attain emancipation and sustainable development. This is because technocrats, civil society organizations and bilateral and multilateral organizations have relegated the importance of inequalities caused by eliticism, bureaucracy, culture, gender, poverty (or class), ethnicity and capitalist economies in colonial inheritor national governments. For this reason, mainstream participatory development has continued to heighten power and market shifts to the already powerful, while limiting access to the rights of participation, economic activity, education, health and water to disadvantaged people.

Ife's deliberative democracy, which is an essential component of participatory democracy, was largely lacking in the operations of technocrats, particularly where they dealt directly with disadvantaged people. In these cases, the binaries of 'uppers' and 'lowers' and 'benefactors' and 'beneficiaries' became the more apparent. These binaries resonate with Pieterse's (2001) dichotomic thinking and are symptomatic of the problems of the assumption that the economically endowed and powerful are rather more able to empower the disadvantaged by spreading benefits of their economic surplus and knowledge to the disadvantaged people (Arnstein, 1971; Ife, 2002).

This research has demonstrated that these power brokers will go to great lengths to ensure a power balance through technologies of participation and partnerships in which affirmative actions fail to work because the disadvantaged must compete as equal partners. In this case, Ife's principles of cooperation, peace and non-violence find less resonance while consensus-building, consciousness-raising and education succumb to the social exclusionary effects of technical expertise and development managerialism because change agents, according to Murunga (2005, p. 10), face "a knowledge challenge ... of thinking and acting … with people … " In the BEIP this challenge arose because technocrats failed to make the critical leap Shivji (1999) recommended to Africa in the words "we have to sit back and ... think with the people, and think for ourselves, for as Wamba-dia-Wamba says: 'people think', and not only think, but also act" (cited in Murunga, 2005, p. 10).

The principles of representation, technical expertise and the conditions of disadvantaged people contributed to a disconnection between policies and practices. Pluralist and elitist approaches that informed participation in the BEIP shared concerns for emancipation and social transformation of the disadvantaged people. However, the idea of facilitating participation, partnerships and empowerment through representatives and technical experts faltered because disadvantaged people could not favourably compete and representatives were not democratically elected. The reproduction of social inequalities and new forms of centralism means that mainstream participatory development co-opts discourses of participation, empowerment and transformation to legitimate opposition while promoting its neoliberal agenda (Cornwall and Brock, 2004).

Since political empowerment cannot be achieved outside of government systems (Brohman, 1996; Friedmann, 1992; Ife, 2002) lest attempts to isolate the disadvantaged people result in ethnic, racial or religious enclaves (Pieterse, 2001), following Ife (2002), this book suggests that the government must deepen participatory approaches in policies and development projects. It has also suggested that balancing between structural and poststructural approaches and between the technical and moral components of development is more likely to achieve positive changes towards emancipation and social transformation. Effort should be made to forge theoretical and practical cooperation and coalition between structures and rights-based approaches with a focus on placing decision-making power, responsibilities and services closer to or with the family/communities. Rather than promulgate national unity by demonizing tribal (or ethnic) identities and affiliations, focus should be on optimizing local potential and celebrating cultural diversity where it enhances wellbeing. Where culture limits wellbeing and access to collective and individual rights, more contextualized strategies should be sought. Such strategies must seek to enable the communities that experience the deleterious effects to evolve their own plans and benefit from the associated processes of learning in tandem.

This means that democratic, cooperative and peaceful, rather than non-democratic, competitive and violent ways of problem-solving are better placed to emancipate and achieve sustainable development (Ife, 2002). Affirmative action policies will go a long way towards realizing this aim where the focus is to increase control of the disadvantaged people over the social, economic and political structures that govern their lives. Governments committed to sustainable development must forge strong partnerships with the civil society and the disadvantaged people. Finally, unless the participation, educational, infrastructural, unemployment, poverty and security issues that inhibit disadvantaged people from optimizing their social, economic and political capitals are addressed, emancipation and sustainable development are unlikely.

Bibliography

African Peer Review Mechanism (APRM). (2006). *Country review report of the Republic of Kenya*. Retrieved March, 2007, from http://www.nepad.org

Ake, C. (1996). *Democracy and development in Africa*. Washington, DC: The Brookings Institution.

Ake, C. (2000). *The feasibility of democracy in Africa*. Dakar: CODESRIA.

Arnstein, S. (1969). A ladder of citizen participation. *Journal of the American Institute of Planners*, 35(4), 216–24.

Arnstein, S. (1971). A ladder of citizen participation. *Journal of American Institute of Planners*, 35(7), 216–24.

Bhabha, H.K. (1994). *The location of culture*. New York: Routledge.

Bhabha, H.K. (1995). Translator translated, interview with W.J.T. Mitchell. *Artforum*, 80–3, 110, 114, 118–19.

Biggs, S., and Smith, G. (1998). Beyond methodologies: Coalition-building for participatory technology development. *World Development*, 26(2), 239–48.

Bopp, M. (1994). The illusive essential: Evaluating participation in non-formal education and community development process. *Convergence*, 27(1), 23–45.

Botchway, K. (2001). Paradox of empowerment: Reflection on a case study from Northern Ghana. *World Development*, 29(1), 135–53.

Brett, E.A. (2003). Participation and accountability in development management. *The Journal of Development Studies*, 40(2), 1–29.

Brohman, J. (1996). *Popular development: Rethinking the theory and practice of development*. Oxford: Blackwell.

Brown, D. (1995). Seeking the consensus: Populist tendencies at the interface between research and consultancy, *AERDD Working Papers*. Reading: University of Reading.

Brown, D. (1998). Professionalism, participation and public good: Issues of arbitration in development management and the critique of the neo-populist approach. In M. Minogue, C. Poldano and D. Hulme (eds), *Beyond the new public management: Changing ideas and practices in governance*. Cheltenham: Edward Elgar.

Brown, D. (2004). Participation in poverty reduction strategies: Democracy strengthened or democracy undermined? In S. Hickey and G. Mohan (eds), *Participation: From tyranny to transformation? Exploring new approaches to participation in development* (pp. 237–51). London: Zed Books.

Buch-Hansen, E. (2002, 21 October). *Are partnerships and participation "magic wands" for promoting sustainability, democratization, equity and poverty reduction?* Paper presented at the Aid Impact Forum seminar on Partnership,

Participation and Empowerment in Development Cooperation – A Role for the Poor – or Legitimizing Devices?, Copenhagen.

Burkey, S. (1993). *People first: A guide to self-reliant, participatory rural development*. London: Zed Books.

Cardoso, F. (1973). Associated-dependent development: Theoretical and practical implications. In A. Stepan (ed.), *Authoritarian Brazil: Origins, policies, and future* (pp. 142–76). New Haven, CT: Yale University Press.

Cardoso, F., and Faletto, E. (1979). *Dependency and development in Latin America*. Berkeley, CA: University of California Press.

Chambers, R. (1974). *Managing rural development: Ideas and experiences from East Africa*. Uppsala: The Scandinavian Institute of African Studies.

Chambers, R. (1983). *Rural development: Putting the last first*. London: Longman.

Chambers, R. (1994a). Participatory rural appraisal (PRA): Analysis and experience. *World Development*, 22(9), 1253–68.

Chambers, R. (1994b). Participatory rural appraisal (PRA): Challenges, potentials and paradigm. *World Development*, 22(10), 1437–54.

Chambers, R. (1995). *Rural Development: putting the last first*. Upper Saddle River, NJ: Prentice Hall.

Chambers, R. (1997). *Whose reality counts? Putting the first last*. London: Intermediate Technology.

Chambers, R. (2002). Relaxed and participatory appraisal: Notes on practical approaches and methods for participants. *PRA/PLA-Related familiarization Workshops*. Sussex: Institute of Development Studies.

Chambers, R. (2005). *Ideas for development*. London: Earthscan.

Choguill, M.B.G. (1996). A ladder of community development for underdeveloped countries. *Habitat International*, 20(3), 431–44.

Cleaver, F. (1999). Paradoxes of participation: Questioning participatory approaches to development. *Journal of International Development*, 11(4), 597–612.

Cook, B. (2003). A new continuity with colonial administration: Participation in development management. *Third World Quarterly*, 24(1), 47–61.

Cook, B. (2004). Rules of thumb for participatory change agents. In S. Hickey and G. Mohan (eds), *Participation: From tyranny to transformation?* (pp. 42–55). London: Zed Books.

Cook, B., and Kothari, U. (eds). (2001). *Participation: The new tyranny?* London: Zed Books.

Cornwall, A. (2000). Beneficiary consumer, citizen: Changing perspectives on participation and poverty reduction. *Sida Studies No. 2*. Stockholm: SIDA.

Cornwall, A. (2002). Making spaces, changing places: Situating participation in development. *IDS Working Paper 170*. Brighton: Institute of Development Studies.

Cornwall, A. (2003). Whose voices? Whose choices? Reflections on gender and participatory development. *World Development*, 31(8), 1325–42.

Cornwall, A. (2004). Spaces for transformation? Reflections on issues of power and difference in participatory development. In S. Hickey and G. Mohan (eds), *Participation: From tyranny to transformation? Exploring new approaches to participation in development* (pp. 75–91). London: Zed Books.

Cornwall, A., and Brock, K. (2004, 20–21 April). *What do buzzwords do for development policy? A critical look at 'poverty reduction', 'participation' and 'empowerment'*. Paper presented at the UNRISD Conference on Social Knowledge and International Policy Making: Exploring the Linkages, Geneva, Switzerland.

Craig, D., and Porter, D. (1997). Framing participation: development projects, professionals and organizations. *Development in Practice*, 7(3), 229–36.

Craig, D., and Porter, D. (2003). Poverty Reduction Strategy papers: A new convergence. *World Development*, 31(1), 53–69.

Eade, D. (ed.) (2003). *Development methods and approaches: Critical reflections*. Oxford: Oxfam UK.

Escobar, A. (1985). Discourse and power in development: Michael Foucault and the relevance of his work to the Third World. *Alternatives*, 10, 377–400.

Escobar, A. (1992). Reflections on development: Grassroots approaches and alternative politics in the Third World. *Futures*, 24, 411–36.

Escobar, A. (1995). Imagining post-development era. In J. Crush (ed.), *The Power of development*. London: Routledge.

Escobar, A. (1996). Imagining a post-development era. In J. Crush (ed.), *The power of development* (pp. 211–27). London: Routledge.

Esteva, G. (1985). Beware of participation, and development: Metaphor, myth, threat. *Development: Seeds of Change*, 3, 77–9.

Feeney, P. (1998). *Accountable aid: Local participation in major projects*. Oxford: Oxfam UK.

Frank, A.G. (1969). *Latin America: Underdevelopment or revolution*. New York: Monthly Review Press.

Freire, P. (1970). *Pedagogy of the oppressed*. New York: Herder and Herder.

Freire, P. (1972). *Pedagogy of the oppressed*. Harmondsworth: Penguin.

Freire, P. (1973). *Education for critical consciousness*. New York: Seabury Press.

Friedmann, J. (1992). *Empowerment: The politics of alternative development*. Oxford: Blackwell.

Furtado, C. (1970). *Economic development of Latin America*. Cambridge: Cambridge University Press.

Gandhi, M. (1964). Gandhi on non-violence: Selected texts from Mahandas K. Gandhi. In T. Merton (ed.), *Non-violence in peace and war*. New York: New Directions.

Gaventa, J. (2002). Introduction: Exploring citizenship, participation and sustainability. *IDS Bulletin*, 33, 1–11.

Gaventa, J. (2004). Towards participatory governance: Assessing the transformative possibilities. In S. Hickey and G. Mohan (eds), *Participation: From tyranny to transformation? Exploring new approaches to participation in development* (pp. 25–41). London: Zed Books.

Gaventa, J., and Valderrama, C. (2001). *Participation, citizenship and local governance in enhancing ownership and sustainability: A resource book on participation*: IFAD, ANGOC and IRR.

Ghai, D. (1988). *Participatory development: Some perspectives from grass-roots experiences*. Geneva: UNRISD.

GOK. (1965). *African socialism and its application to planning in Kenya*. Nairobi: Government Printers.

GOK. (1976). *Report of the National Committee on Educational Objectives and Policies: The Gachathi report*. Nairobi: Ministry of Education.

GOK. (2001a). *Education for all in Kenya: A national handbook on EFA 2000 and beyond; Meeting our collective commitments to Kenyans*. Nairobi: Government Printer.

GOK. (2001b). *Poverty reduction strategy paper (PRSP) for the period 2001–2004*. Nairobi: Government Printer.

GOK. (2002a). *Action plan for the implementation of the poverty reduction strategy paper*. Nairobi: Government Printer.

GOK. (2002b). *Support for increasing access and equity in the provision of quality basic education: The OPEC fund for international development*. Nairobi: Government Printer.

GOK. (2003a). *Basic Education Improvement Project Training Module*. Nairobi: Ministry of Education, Science and Technology.

GOK. (2003b). *Manual for implementation of Basic Education Improvement Project: GOK/OPEC Fund*. Nairobi: Unpublished.

GOK. (2003c). *Public expenditure review*. Nairobi: Government Printer.

GOK. (2003d). *Report of the National Conference on Education and Training*. Nairobi: Government Printer.

GOK. (2003e). *Report of the sector review and development direction*. Nairobi: Ministry of Education, Science and Technology.

GOK. (2004a). Economic recovery strategy for wealth and employment creation. Nairobi: Government Printer.

GOK. (2004b). *Investment programme for the economic recovery strategy for wealth and employment creation 2003–2007*. Nairobi: Noel Creative Media.

GOK. (2005a). *Kenya education sector support programme 2005–2010: Delivering quality education and training to all Kenyans*. Nairobi: Government Printer.

GOK. (2005b). *Sessional paper No. 1 of 2005: A policy framework for education, training and research*. Nairobi: Government Printer.

GOK. (2007). The ERS, mid-term review, popular version. In *Ministry of Planning and National Development – Monitoring and Evaluation* (ed.). Nairobi: Media Edge Interactive.

Gregory, B. (2007). Bringing back the buck: Responsibility and accountability in politics and the state sector. *Public Sector*, 30(2), 2–6.

Harper, C. (1997). Using grassroots experience to inform macro level policy: An NGO perspective. *Journal of International Development*, 9(5), 771–8.

Hayward, C., Simpson, L., and Wood, L. (2004). Still left out in the cold: Problematising participatory research and development. *Sociologia Ruralis*, 44(1), 95–108.

Hickey, S. (2002). Transnational NGDOS and participatory forms of rights-based development: Converging with the local politics of citizenship in Cameroon. *Journal of International Development*, 14, 841–57.

Hickey, S., and Mohan, G. (2004a). Towards participation as transformation: Critical themes as challenges. In S. Hickey and G. Mohan (eds), *Participation: From tyranny to transformation? Exploring new approaches to participation in development* (pp. 3–24). London: Zed Books.

Hickey, S., and Mohan, G. (eds). (2004b). *Participation: From tyranny to transformation? Exploring new approaches to participation in development*. London: Zed Books.

Hickey, S., and Mohan, G. (2005). Relocating participation within a radical politics of development. *Development and Change*, 36(2), 237–62.

Ife, J. (2002). *Community development: Community-based alternatives in an age of globalization* (2nd edn). Frenchs Forest, NSW: Pearson Education Australia.

Isbister, J. (1991). *Promises never kept: The betrayal of social change in the third world*. West Hartford, CT: Kumarian Press.

Kapoor, I. (2002a). Capitalism, culture, agency: Dependency versus postcolonial theory. *Third World Quarterly*, 23(4), 647–64.

Kapoor, I. (2002b). The devil's in the theory: A critical assessment of Robert Chambers' work on participatory development. *Third World Quarterly*, 23(1), 101–17.

Kapoor, I. (2005). Participatory development, complicity and desire. *Third World Quarterly*, 26(8), 1203–1220.

Kipngetich, J. (2001). *Devolution of power*. Retrieved October 21, 2007, from www.commonlii.org/ke/other/KDKRC/2001/13.html

Klees, S.J. (2001). World Bank development policy: A SAP in SWAPs clothing. *Current Issues in Comparative Education*, 3(2), 110–21.

Kothari, R. (1988). *Rethinking development: In search of humane alternatives*. Delhi: Ajanta.

Latouche, S. (1993). *In the wake of the affluent society: An exploration of post-development*. London: Zed Books.

Lewis, W.A. (1964). Economic development with unlimited supplies of labour. In S.P. Singh (ed.), *The economics of underdevelopment* (pp. 400–49). London: OUP [published originally in 1954 in The Manchester School of Economic and Social Studies 22:2].

Makuwira, J. (2003). *Non-Governmental Organizations (NGOs) and Participatory Development in Basic Education in Malawi*. Unpublished PhD Thesis, University of New England, Queensland.

Makuwira, J. (2006). Development? Freedom? Whose development and Freedom? *Development in Practice*, 16(2), 193–200.

Manzo, K. (1991). Modernist discourse and the crisis of development theory. *Studies in Comparative International Development*, 26(2), 3–36.

Marden, D., and Oakley, P. (eds). (1990). *Evaluating social development projects. Development guidelines No. 5.* Oxford: Oxfam UK.

Martinussen, J. (1997). *Society, state and market: A guide to competing theories of development*. London: Zed Books.

Mbaku, J.M. (2000a). *Bureaucratic and political corruption in Africa: The public choice perspective*. Malabar, FL: Krieger.

Mbaku, J.M. (2000b). Governance, wealth creation and development in Africa: The challenges and the prospects. *African Studies Quarterly: The Online Journal for African Studies*, 4(2), 1–18.

Mbaku, J.M. (2004). NEPAD and prospects for development in Africa. *International Studies*, 41(4), 386–409.

Michener, V.J. (1998). The participatory approach: contradictions and co-option in Burkina Faso. *World Development*, 26(12), 2105–18.

Mikkelsen, B. (2005). *Methods for development work and research: A new guide for practitioners* (2nd edn). New Delhi: Sage.

Mohan, G., and Holland, J. (2001). Human rights and development in Africa: Moral intrusion or empowering opportunity? *Review of African Political Economy*, 28(88), 177–96.

Mohan, G., and Stokke, K. (2000). Participatory development and empowerment: The dangers of localism. *Third World Quarterly*, 21(2), 247–69.

Mukudi, E. (1999). Public funding of primary education in Kenya: Recent trends, challenges, and implications for the future. *International Journal of Educational Reform*, 8(4), 383–7.

Mukudi, E. (2004). The effects of user-fee policy on attendance rates among Kenyan elementary school children. *International Review of Education*, 50(5–6), 447–61.

Mulenga, D. (1999). Reflections of the practice of participatory research in Africa. *Convergence*, 32(1–4), 33–43.

Mulwa, F.W. (1994). *Enabling the rural poor through participation*. Eldoret, Kenya: AMECEA Gaba.

Murunga, G.R. (2002). A critical look at Kenya's non-transition to democracy. *Journal of Third World Studies*, 19(2), 89–111.

Murunga, G.R. (2005). A note on the knowledge question in Africa's development. *CODESRIA Bulletin*, 3 and 4, 8–10.

Murunga, G.R., and Nasong'o, S.W. (eds). (2007). *Kenya: The struggle to democracy*. London: Zed Books.

Nasong'o, S.W. (2004). From political dictatorship to authoritarian economism: Plural politics and free markets reforms in Africa. *Journal of Third World Studies*, 21(2), 107–25.

Ndengwa, S. (1996). *The two faces of civil society*. West Hartford, CT: Kumarian Press.

Nkinyangi, J.A. (1981). Education for nomadic pastoralists: Development planning by trial and error. In J.G. Galaty, D. Aronson and P.C. Salzman (eds), *The future of pastoral peoples* (pp. 183–96). Ottawa: International Research Development Centre.

Peet, R., and Hartwick, E. (1999). *Theories of development*. New York: The Guilford Press.

Pieterse, J.N. (2000). After post-development. *Third World Quarterly*, 21(2), 175–91.

Pieterse, J.N. (2001). *Development theory: Deconstructions/reconstructions*. London: Sage.

Pieterse, J.N. (2002). Global inequality: Bringing politics back in. *Third World Quarterly*, 23(6), 1023–46.

Pretty, J. (1995). Participatory learning for sustainable agriculture. *World Development*, 23(8), 1247–63.

Pretty, J., and Guijt, I. (1992). Primary environmental care: An alternative paradigm for development assistance. *Environment and Urbanization*, 4(1), 23.

Rahnema, M. (1992). Participation. In W. Sachs (ed.), *The development dictionary*. London: Zed Books.

Rahnema, M. (1997). Towards post-development: Searching for signposts, a new language and new paradigm. In M. Rahnema and V. Bawtree (eds), *The post-development reader* (pp. 377–404). London: Zed Books.

Rowlands, J. (1997). *Questioning empowerment: Working with women in Honduras*. Oxford: Oxfam UK.

Rowlands, J. (1998). A word of our time, but what does it mean? Empowerment in the discourse and practice of development. In H. Afshar (ed.), *Women and empowerment: Illustrations from the Third World* (pp. 11–34). Basingstoke: Macmillan.

Sachs, W. (ed.). (1992). *The development dictionary: A guide to knowledge as power*. London: Zed Books.

Saitoti, G. (2002). *The challenges of economic and institutional reforms in Africa*. Aldershot: Ashgate.

Seers, D. (1972). What are we trying to measure? *Journal of Development Studies*, 8(3), 22–36.

Sen, A. (1999). *Development as freedom*. Oxford: Oxford University Press.

Shepherd, A. (1998). *Sustainable development*. Basingstoke: Macmillan.

Shivji, I. (1999). Who are the poor and whose justice are they accessing? The dilemmas of a legal aid activist. In H. Othman and C.M. Peter (eds), *Perspectives on legal aid and access to justice in Zanzibar* (pp. 5–16). Zanzibar: Zanzibar Legal Services Centre.

Shivji, I. (2003). The struggle for democracy: http://www.marxists.org/subject/africa/shivji/struggle-democracy.htm.

Sifuna, D.N. (2005a). *The illusion of universal free primary education in Kenya*. Retrieved January, 15, 2007, from http//africa.peacelink.org/wajibu/articles/art_6901.html

Sifuna, D.N. (2005b). Increasing access and participation of pastoralist communities in primary education in Kenya. *International Review of Education*, 51(5–6), 499–516.

Simon, D. (1999). Development revisited: Thinking about, practising and teaching development after the Cold War. In D. Simon and A. Narman (eds), *Development as theory and practice* (pp. 17–54). Harlow: Addison Wesley Longman.

Spivak, G. (1985). The Rani of Simur: An essay in the reading of archives. *History and Theory*, 24(3), 247–72.

Thomas, B.P. (1987). Development through harambee: Who wins and who loses? Rural self-help projects in Kenya. *World Development*, 15(4), 463–81.

Tondon, Y. (1995). Poverty, processes of impoverishment and empowerment: Review of current thinking and action. In N.C. Singh and V. Titi (eds), *Empowerment: Towards sustainable development* (pp. 31–7). Atlantic Highlands, NJ: Zed Books.

Tritter, J.Q., and McCallum, A. (2006). The snakes and ladders of user involvement: Moving beyond Arnstein. *Health Policy*, 76, 156–68.

UNESCO (2006). Draft of strategy of education for sustainable development in Sub-Saharan Africa. Nairobi: UNESCO.

Visvanathan, S. (1988). On the annals of the laboratory state. In A. Nandy (ed.), *Science, hegemony and violence: A requiem for modernity*. New Delhi: Oxford University Press.

White, S.C. (1996). Depoliticizing development: The uses and abuses of participation. *Development in Practice*, 6(1), 6–15.

Williams, G. (2004). Evaluating participatory development: tyranny, power and (re)politicisation. *Third World Quarterly*, 25(3), 557–78.

Willis, K. (2005). *Theories and practices of development: Routledge perspectives on development*. London: Routledge (UK).

World Bank (1997). *World Bank Development Report. The state in a changing world*. Oxford: Oxford University Press.

Yamamori, T., Myers, B.L., Bediako, K., and Reed, L. (eds). (1996). *Serving with the poor in Africa*. Monrovia and California: MARC.

Index